OVERKILL

OVERKILL

WHEN MODERN MEDICINE GOES TOO FAR

Paul A. Offit, MD

HARPER
An Imprint of HarperCollins*Publishers*

This book is written as a source of information only. The information contained in this book should by no means be considered a substitute for the advice of a qualified medical professional. All efforts have been made to ensure the accuracy of the information contained in this book as of the date published. The author and the publisher expressly disclaim responsibility for any adverse effects arising from the use or application of the information contained herein.

HarperCollins books may be purchased for educational, business, or sales promotional use. For information, please email the Special Markets Department at SPsales@harpercollins.com.

FIRST EDITION

Designed by Elina Cohen

Library of Congress Cataloging-in-Publication Data has been applied for.

ISBN 978-0-06-294749-9

20 21 22 23 24 LSC 10 9 8 7 6 5 4 3 2 1

It ain't what you don't know that gets you into trouble. It's what you know for sure that just ain't so.

—MARK TWAIN

Contents

Part IV: Cancer Screening

Part V: Surgery

Part VI: Common Beliefs

OVERKILL

Prologue: Ignoring James Lind

In 1740, George Anson, the First Lord of the Admiralty of the British Royal Navy, sailed around the world. When he launched, 961 sailors climbed aboard his ships. When he finished, 335 got off. The rest had died from scurvy.

Anson's expedition wasn't unique. Scurvy, a disease known since ancient times, often afflicted sailors on long journeys. Symptoms included bleeding gums, ready bruising, hemorrhages, swollen legs, weakness, apathy, shortness of breath, fatigue, joint pain, slow-healing wounds, and eventually death from heart disease.

Although everyone knew about scurvy, no one knew what caused it or how to treat it. Seven years after Anson's expedition, all that would change. Or, at least, it should have changed. Although more than a century would pass before scientists proved that scurvy is caused by a deficiency of vitamin C, one man figured out how to treat it before vitamins were discovered. Unfortunately, his findings were ignored.

In March 1747, a young Scottish surgeon named James Lind climbed aboard the HMS *Salisbury*, a fifty-gun ship charged with blocking the English Channel. At the time, Britain was embroiled in the War of the Austrian Succession against France and Spain. Two months after leaving port, with many sailors belowdecks suffering from scurvy, Lind performed what many consider the first clinical study. He split twelve sailors into six groups of two men, offering each group one of the following "remedies": a quart of cider every day; two spoonfuls of vinegar three times a day; half a pint of seawater every day; twenty-five drops of elixir of vitriol (i.e., diluted sulfuric acid) three times a day; a paste made of garlic, mustard seed, dried radish root, and gum myrrh three times a day; or two lemons and one orange every day.

Six years later, in 1753, Lind published his results. Titled *Treatise of the Scurvy,* it was dedicated to Lord Anson. Only one group had benefited. "The most sudden and visible good effects were perceived from the use of oranges and lemons," he wrote. "One of those who had taken them, being at the end of six days fit for duty. The other . . . was appointed to attend to the rest of the sick." Although Lind didn't know why citrus fruits had cured scurvy, he clearly had shown that they had.

No one paid attention to Lind's breakthrough observation. Doctors continued to treat scurvy with elixir of vitriol, even though Lind had shown it was useless. During the next six years, among the 176,000 sailors in the British fleet, 18,500 died from scurvy—all because doctors had continued to treat them with medicines that had no chance of working. Indeed, more sailors died from scurvy than were killed in battle.

In 1794, about forty years after James Lind's publication, Rear Admiral Alan Gardner insisted on bringing large quantities of lemons aboard a twenty-three-week nonstop voyage to India. No one got scurvy. The British Admiralty could ignore Lind's study no longer. In May 1796, the Sick and Hurt Board in Britain, which recently had added two naval surgeons well acquainted with Lind's findings, agreed to supply all naval ships on foreign service with lemon juice. Three years later, the board extended its recommendation to all British ships. Scurvy disappeared from the British Royal Navy.

Because the navy could now block French ports for years at a time, sea power won the Napoleonic Wars. "Of all the means which defeated Napoleon," noted one observer, "lemon juice and the carronade gun were the two most important."

Why did it take so long for doctors to accept James Lind's findings? At the time, health officials estimated that a cure for scurvy would have doubled the efficiency of the British fleet. Surely, no admiral would have turned down such a prospect. So, why did they? Given that this was one of the most remarkable examples of medical denialism in history, researchers have offered several theories:

Theory 1: The description of Lind's experiment on the HMS *Salisbury* was only 4 pages long, a problem when the *Treatise* itself was 450 pages long. It's easy to see how Lind's critical finding could have been missed.

Theory 2: Lind knew that lemons and oranges worked; he just didn't know why. He assumed that the fruits were treating a digestive disorder, not a vitamin deficiency. In fairness, it wasn't until 1912, when English biochemist Frederick Hopkins published his paper on "accessory food factors," that scientists had any idea what vitamins were or that they could be obtained only from food. (Of interest, humans are among a group of only a few animals, including guinea pigs and monkeys, that can't synthesize their own vitamin C. That's why sailors who ate rats on the ship were less likely to get scurvy: rats are a rich source of the vitamin.)

Theory 3: Lind was a poor advocate for his findings. "The province has been mine to deliver precepts," he wrote. "The power is in others to execute." Trusting the British Admiralty to be immediately compelled by his scientific study was, at the very least, naïve.

Theory 4: It's hard to counter established treatments and biases. "It is no easy matter to root out old prejudices," wrote Lind, "or to overturn opinions established by time, custom, and great authorities."

James Lind died on July 13, 1794, one year before the Admiralty required citrus fruits for British ships. Although he may have been ignored in his time, Lind's work has been immortalized on the official crest of the British Institute of Naval Medicine, which depicts a lemon tree sitting atop an ocean.

TODAY, we look back at the story of James Lind and shake our heads. Surely, when confronted with clear, unambiguous evidence that a treatment works or doesn't work, we embrace the findings. What most people would be surprised to learn, however, is that, in far too many instances, we do not. In the pages that follow, I will describe situations in which clinicians have ignored a wealth of evidence and continued to prescribe medicines, or perform surgeries, or promote cancer screening programs, that have been shown to do more harm than good. Sadly, in certain situations, we are still figuratively offering "elixir of vitriol," treatments or preventives that have been shown again and again to be at best useless and at worst dangerous.

In the epilogue, we'll talk about why this is happening and what can be done to stop it. The good news is that there's a way out of this.

THROUGHOUT this book, many studies will be described. Such reading can be daunting or, worse, boring. But if you're going to be convinced that certain well-established, accepted practices might be wrong, the data must do the convincing. Otherwise, I'd just be asking you to trust me. And while medical gurus such as Mehmet Oz and Deepak Chopra or celebrities such as Jenny McCarthy and Gwyneth Paltrow can probably get away with that, I can't. In the end, you shouldn't trust me; you should trust high-quality, reproducible scientific studies that are performed in well-respected academic centers and published in prestigious medical journals.

PART I
Infections

1

Treating Fever Can Prolong or Worsen Illness

Question: Why is it that every warm- and cold-blooded animal that has walked, flown, swum, or crawled on the face of the earth for the past 600 million years has the capacity to make fever? Isn't it possible that fever is an adaptive response to the environment, allowing us to survive, rather than a maladaptive response causing unnecessary suffering? And isn't it then possible that reducing fever with medicines such as aspirin, acetaminophen (Tylenol), or ibuprofen (Motrin, Advil) might do more harm than good?

On November 6, 2015, the short-lived Cinemax series *The Knick* aired an episode titled "Wonderful Surprises." Promoted with the tagline "Modern medicine had to start somewhere," *The Knick* centers on a fictional New York City hospital in the early 1900s. The main character, played by Clive Owen, is Dr. John W. Thackery, an innovative, arrogant, cocaine-addicted surgeon. In "Wonderful Surprises," Thackery is confronted by his past when his ex-girlfriend, Abigail Alford, played by Jennifer Ferrin, is admitted to the hospital with seizures caused by syphilis. Today syphilis is treated with penicillin; but this was 1906. Penicillin hadn't been discovered yet. So Thackery tries something that shocked the viewing audience.

Working in his makeshift laboratory, Thackery finds that he can kill syphilis bacteria in a petri dish with heat. Then he finds that if he raises the body temperature of a pig he has infected with syphilis to 107 degrees, the pig will completely recover. Secure in the knowledge that heat is curative, Thackery decides that the best way to increase Abigail's temperature is to inject malaria parasites into her bloodstream. Thackery knows that malaria causes high, unrelenting fevers; he also knows that he can treat malaria with quinine, which has been available since the mid-1800s. When Abigail's temperature rises, Thackery checks her blood to see if the syphilis bacteria have disappeared. They haven't. So, he takes the next step. Over the objections of a fellow surgeon, who screams, "You're frying her brain!" Thackery puts Abigail in a fever cabinet, a device that resembles an oven. It raises her temperature even further. After a few days in the fever cabinet, and still infected with malaria, Abigail sees her syphilis cured. Jubilant, the surgeons throw her into a bathtub, pour ice water over her entire body, and treat her with quinine.

The medical advisor to *The Knick* was Dr. Stanley Burns, founder of the Burns Archive, the world's largest collection of turn-of-the-century medical photographs. As a medical historian, Burns was obsessed with the notion that the medicine practiced on *The Knick* be realistic. He insisted that the antiseptic atomizers used in the operating rooms, the crude X-ray machines used by radiologists, and the prosthetic finger worn by a recurring character all be exactly as they would have appeared in the early 1900s. This attention to detail even extended to a scene in which an older surgeon (played by Ben Livingston), while flirting with a nurse during a surgical procedure, sets himself on fire and dies immediately. Ether, which was commonly used as an anesthetic, is highly flammable. When the surgeon cauterizes a wound near the ether mask, the spark initiates a fire that kills him.

The story of the spontaneously combusting surgeon is entirely plausible. But what about the malaria injections and the fever cabinet? Are they plausible? Did doctors really induce high fevers to treat syphilis? In fact, beginning in 1917, they did. A decade later, the process of curing syphilis by injecting malaria parasites won a Nobel Prize for its inventor.

WHILE working at the Lunatic Asylum of Lower Austria in 1883, Julius Wagner-Jauregg noticed that a woman with severe psychosis caused by syphilis was cured after surviving a strep infection that had caused a high fever. Wagner-Jauregg was convinced that fever could be used to treat the seizures, dementia, delusions, and paralysis of syphilis. In 1887, he published an article titled "The Effect of Feverish Disease on Psychosis." Wagner-Jauregg intentionally injected strep bacteria into the bloodstreams of patients with syphilis. Although many developed fevers, he was disappointed by the results: the fevers weren't high enough, and the cures were inconsistent. So he tried something else.

In 1917, Wagner-Jauregg received a message from a colleague, Dr. Alfred Fuchs, informing him of a soldier who had recently contracted malaria. He pleaded with Fuchs not to treat the soldier until he could take a sample of the man's blood. After securing the blood, which was teeming with malaria parasites, Wagner-Jauregg injected it into nine patients suffering from paralysis caused by syphilis infection of the brain. After they had endured high, spiking fevers for more than a week, he treated them with quinine. One patient died, two were sent to sanatoriums, and six improved dramatically. In 1921, Wagner-Jauregg published an article reporting the results of more than two hundred additional patients he had treated, fifty of whom had recovered sufficiently enough to go back to work.

In 1927, Wagner-Jauregg won the Nobel Prize in Medicine "for his discovery of the therapeutic value of malaria inoculation in the treatment of dementia paralytica." Remarkably, malaria therapy for syphilis extended into the 1950s, when penicillin finally became widely available.

Although Julius Wagner-Jauregg was the first to use fever to treat patients with severe and often fatal infections, he wasn't the first to recognize fever's importance. The concept of fever as a valuable tool to fight infections is 2,500 years old. Hippocrates believed that disease was caused when one of the four humors (black bile, yellow bile, blood, and phlegm) was produced in excess. Fever, according to Hippocrates, cooked the raw humor, leading to healing. Although he wasn't right about the four humors, Hippocrates recognized that fever was an important response to infection, not a malevolent bystander.

Julius Wagner-Jauregg had proved that fever was valuable. What he hadn't shown, however, was that reducing fever could be harmful. Fifty years would pass before those studies would be performed, and when they were, researchers found something that modern medicine has consistently chosen to ignore.

Fever-reducing Medicines Prolong and Worsen Infections in Experimental Animals

In 1975, Matthew Kluger, working in the Department of Physiology at the University of Michigan Medical School, performed a breakthrough experiment . . . on lizards. Kluger injected 140 lizards with *Aeromonas hydrophila,* a bacterium that he knew could kill them. Then he put the lizards into chambers set at different temperatures:

At 93 degrees, 0 percent of the animals survived.
At 97 degrees (normal lizard temperature), 25 percent survived.
At 104 degrees, 75 percent survived!

Kluger then wondered what would happen if, after infecting the lizards, he treated them with fever-reducing medicines (antipyretics). (Aspirin, acetaminophen [Tylenol], and ibuprofen [Motrin or Advil] are all antipyretics.) Lizards that were allowed to increase their body temperature survived the week. Those treated with antipyretics were dead within three days.

During the next forty years, researchers performed experiments on a variety of animals infected with a variety of viruses or bacteria that were either treated or not treated with antipyretics. They found that antipyretics decreased survival in ferrets infected with influenza; dogs infected with herpes; goldfish and sockeye salmon infected with *Aeromonas*; mice infected with coxsackievirus, polio, or herpes viruses; and rabbits infected with pneumococcus, staphylococcus, *Escherichia coli*, pseudomonas, streptococcus, *Pasteurella*, myxovirus, or vaccinia virus. In every study, antipyretics worsened illness and increased mortality. In *every study*. No study in experimental animals has ever shown that reducing fever shortened the course of illness.

Then investigators studied people.

Fever-reducing Medicines Prolong and Worsen Infections in People

What follows is a list of seven studies and one review of forty-two more studies.

In 1975, researchers at the University of Illinois College of Medicine studied forty-five young adults experimentally infected with a common cold virus (rhinovirus). Half the subjects were treated with aspirin, and half weren't. Those treated with aspirin shed virus from their noses significantly longer than those who weren't treated.

In 1989, researchers at Johns Hopkins Hospital studied seventy children under twelve years of age who were suffering from chicken pox (varicella). Children were given acetaminophen (Tylenol) or a placebo. The time to fully heal the chicken pox blisters was significantly longer in the Tylenol group.

In 1990, researchers at the University of Adelaide in South Australia inoculated sixty healthy adult volunteers with rhinovirus. Volunteers were then given aspirin, acetaminophen, ibuprofen, or a placebo. Those treated with antipyretics had a markedly reduced immune response to the virus as well as worse congestion, runny nose, sneezing, sore throat, swollen lymph nodes, and cough. Ironically, the study was supported in part by McNeil Consumer Products, the makers of Tylenol.

In 1992, researchers at the University of Helsinki in Finland studied one hundred children with salmonella infection and found that those without fever shed bacteria in their stool for twelve days, whereas those with high fevers shed bacteria for only two days. Again, fever had reduced the duration of the infection and hastened recovery.

In 1994, researchers at the Fujimoto Children's Hospital in Japan studied two hundred children who had fever caused by viral infections of the nose and throat. Half the patients received frequent doses of acetaminophen, and half didn't. Those given acetaminophen were significantly more likely to develop severe pneumonia requiring hospitalization.

In 2000, researchers at the University of Maryland School of Medicine studied people experimentally infected with influenza virus or shigella (an intestinal bacterial infection). Half the subjects were treated with aspirin or acetaminophen, and half weren't. Subjects who were treated with antipyretics had illnesses that were more severe and lasted three to four days longer.

In 2005, researchers at the University of Miami School of Medicine studied eighty patients admitted to the trauma intensive care unit. Half were treated aggressively with acetaminophen whenever they developed fever, and half weren't. (Inflammation following severe trauma can itself cause fever.) Patients whose initial fevers were treated with antipyretics were more likely to develop infections and more likely to die from those infections than those whose fevers weren't treated.

In 2016, researchers in Hungary summarized the results of forty-two studies that had examined the relationship between fever and survival in patients with severe bloodstream infections (sepsis). They found that the death rate was 22 percent in patients with fever,

31 percent in patients with normal body temperature, and 47 percent in patients with temperatures lower than normal. Again, fever increased survival.

WHY does fever lessen the severity of infections? One explanation is that some bacteria (such as the one that causes syphilis) are killed at higher temperatures. Most bacteria, viruses, fungi, and parasites, however, aren't susceptible to these higher temperatures. The reason that fever ameliorates infections is that *our immune system works better at higher temperatures*—much better. In order to understand why this is true, we need to understand how our immune system works, how our body generates fever, how antipyretics lessen fever, and why giving antipyretics routinely is arguably one of the most unnatural and ill-advised things we do in modern medicine.

The immune system can be divided into two parts: primitive and adaptive. The primitive part of our immune system has been around for at least 250,000 years. When bacteria are inadvertently injected under our skin or enter our bloodstream, certain types of primitive immune cells known as macrophages and dendritic cells spring into action. The first thing these cells do is make several proteins called cytokines, which travel to an area deep in our brain called the hypothalamus. The hypothalamus acts as a thermostat, setting the desired body temperature. Once they enter the hypothalamus, these cytokines stimulate the production of another protein that is critical to our survival. It's called prostaglandin E2, or PGE2. Typically, the hypothalamus's thermostat is set to center around 98.6 degrees Fahrenheit, with temperatures ranging from 96.5 to 99.5 degrees depending on the time of day. But when PGE2 is produced, the thermostat is reset to center around a higher temperature, like 104 degrees.

Now, in the face of this resetting, we seek heat, desperately. We shiver to generate heat. We shunt blood away from our arms and legs and into our core to prevent heat loss. We put on heavier clothing and climb under the covers. We put logs on the fire and hot water bottles under the bedsheets. Despite doing all these things, we still feel cold because our body is telling us that we aren't hot enough. We need to get warmer and warmer.

Now, at this higher temperature, the immune cells at the center of our primitive immune system (called neutrophils) are at peak performance. Neutrophils ingest bacteria and kill them. (Pus is comprised almost entirely of neutrophils.) What most people don't realize is that neutrophils travel to the sites of infection faster and ingest and kill bacteria better at higher temperatures. That's why antipyretics prolong or worsen serious bacterial infections such as sepsis and pneumonia. When *antibiotics* bring down fever, it's because the bacteria that are causing the disease are being eliminated. So, inflammation and consequent fever subside. When *antipyretics* bring down fever, however, it's because the medication is subverting the body's natural physiological response to enhance the immune response. People may feel better when antipyretics lessen fever, but these drugs do nothing to actually lessen infection. Quite the opposite.

The second part of our immune system, which appeared in humans more recently during evolution, is adaptive. This is the part of the immune system that learns from each infection in order to prevent the next one. For example, within a couple of weeks of being infected with a virus (such as rhinovirus or influenza), we make an immune response *specific* for that virus. (Neutrophils, however, are indiscriminate. They'll kill any invading bacteria.) One type of adaptive immune cell is called a B cell. B cells make antibodies that are specific for harmful invaders and help to neutralize them. Like neutrophils, B cells work better and faster at higher temperatures. That's why the volunteers who were experimentally inoculated with a common cold virus had blunted virus-specific antibody responses after receiving antipyretics. Their adaptive immune response wasn't working as well.

By blocking PGE2, antipyretics interfere with our attempt to raise our internal thermostat so that our primitive and adaptive immune systems can work more efficiently. It is, at the very least, an act of hubris on our part to counter this natural, adaptive, lifesaving process.

Probably nothing demonstrates the importance of fever in generating adaptive immune responses better than what happens when we give antipyretics either before or immediately after vaccination. In 2009, researchers in the Czech Republic divided 460 children who were about to receive vaccines into two groups: one group received acetaminophen every six to eight hours for twenty-four hours; the other didn't. Investigators found a significant decrease in antibody responses to vaccines against pneumococcus, *Haemophilus influenzae* type b (Hib), diphtheria, tetanus, and pertussis in those who had received the anti-fever medicine. The authors concluded that "prophylactic administration of antipyretic drugs at the time of vaccination should not be routinely recommended."

In 2018, researchers in Australia evaluated 3,300 children who had received an influenza vaccine to determine whether antipyretics influenced immune responses. Again, antipyretics significantly blunted the children's antibody response to the vaccine. These studies, however, have had little to no impact on this practice.

The price of treating fevers can also be seen at the population level. Researchers in Canada found that people who used antipyretics during an influenza epidemic were more likely to shed virus for longer periods and, presumably because they felt better, more likely to leave their homes and spread their illnesses. As a consequence, societies that use antipyretics more frequently suffer a 5 percent increase in the number of influenza cases and deaths. Researchers have estimated that avoiding antipyretics during influenza season would save about seven hundred lives in the United States and about forty thousand lives worldwide every year.

To sum up: Antipyretics have been shown to decrease survival in experimental animals, prolong illness or decrease survival in people, and decrease the immune response to vaccines. Indeed, fever is so universally required to help fight infections that even plants have been found to increase their temperatures by about three degrees when experimentally infected with a fungus.

Given all this, why do we continue to insist on crippling our immune systems with antipyretics?

Three Reasons We Insist on Using Antipyretics

1. To Prevent a "Fried Brain"

When *The Knick*'s Dr. John Thackery puts Abigail in the fever cabinet to treat her syphilis, his colleague, who has become increasingly agitated as her temperature rises to unimaginable heights, screams, "You're frying her brain!" Thackery's colleague is expressing a commonly held belief; specifically, that high temperatures can cause permanent brain damage. In a sense, he is right.

People experience two different types of fever: one that comes from inside the body and one from outside. The first is natural, or physiological, fever. When we are infected with bacteria or viruses, we adjust our internal thermostat to a higher temperature. Assuming that the germ that is causing the infection doesn't infect the brain (encephalitis) or the lining of the brain (meningitis), natural fevers don't cause brain damage.

However, when Dr. Thackery puts Abigail in the fever cabinet, he isn't inducing a natural fever. He is overriding her internal thermostat and causing an unnatural, nonphysiological, environmental fever. Although people are no longer placed in fever cabinets, we experience environmental fevers all the time. For example, otherwise healthy athletes, outdoor laborers, and military personnel engaged in strenuous activities on hot, humid days can occasionally experience something called heatstroke.

Typically, our bodies dissipate heat by sweating. As the sweat evaporates, our bodies cool. When the humidity is high, however, evaporation doesn't occur. When temperatures get high enough, people can suffer confusion, headache, and dizziness. If temperatures continue to rise, these people can suffer muscle breakdown, kidney failure, multi-organ failure, and death. Indeed, heatstroke causes about six hundred deaths every year in the United States.

In India, during a severe heat wave in 2015, more than twenty-five hundred people died.

Heatstroke can also affect young children left in cars on a hot day. Between 1998 and 2011, about five hundred children died in cars when the windows were closed and the child was unattended. Most were under two years old. In these situations, the temperature in the cars could exceed 120 degrees Fahrenheit. High temperatures caused by environmental factors can kill. In other words, Thackery's colleague is right to worry about the fever cabinet.

Whereas environmental fevers can cause harm, natural fevers (such as those induced by infections) can't. So, fevers caused by the body's response to an infection aren't going to "fry your brain."

2. To Prevent Febrile Seizures

The most common type of seizure in young children is caused by fever. Every year, about 5 percent of children between six months and five years of age suffer a "febrile seizure." These seizures are brief, usually lasting less than thirty seconds. The reason that young children are prone to seizures with fever isn't clear. But three things are clear: First, febrile seizures are benign; they're not a form of epilepsy. Children who have suffered febrile seizures are indistinguishable developmentally from those who have never suffered them. Second, about one-third of children with febrile seizures will experience a recurrence. Third, because febrile seizures can recur, people reasonably assume that antipyretics given at the first sign of fever can prevent them. The problem with this strategy is that it doesn't work. A recent review of twenty-six studies involving twenty-five hundred children who either did or didn't receive antipyretics at the first sign of fever found that antipyretics didn't prevent febrile seizures.

So, although febrile seizures are difficult to watch, there really isn't anything you can do to prevent them. More important, children grow out of them, and they don't cause long-term problems.

3. To Feel Better

The main reason that we treat fever is to feel better. We do it so that our hearts no longer beat faster. We do it so we don't have to shiver. And we do it so that we no longer have to suffer the side effects of fever, such as headache, muscle aches, and fatigue.

But the simple truth is that we're not *supposed* to feel better. We're supposed to stay in bed and let our body fight the infection the best way it can: at a higher temperature, a temperature that allows every aspect of our immune system to work better. The trade that we make for temporarily feeling better is that mild infections might last longer and severe infections might become more severe—or fatal.

Pharmaceutical companies haven't helped. Ever since antipyretics first became available (aspirin in 1899, acetaminophen in 1948, and ibuprofen in 1961), pharmaceutical companies have been promoting them with ads such as "Let's get that temperature down," "Just what the doctor ordered," and "When fever and aches have little Tyler corralled—Tylenol." We are constantly bombarded with the notion that fever is bad and that it must be reduced or eliminated.

CHANGING the culture of treating fever won't be easy, but it is possible. The following story shows how.

In 2018, a teenage girl was admitted to the hospital with fever and hip pain. One week earlier, she had been struck in the hip during a soccer match. An MRI showed that she had an infection of her hip joint, hip bone, and hip muscles. Also, the vein around the joint had a clot that had become infected. The bacterium that was causing this infection was staph (MRSA), which was also found in her bloodstream. She was immediately given an intravenous antibiotic that should have killed the MRSA. Nonetheless, day after day, the MRSA continued to grow in her bloodstream. Eventually, the bacteria spread to other joints and bones. It also spread to her lungs and brain, causing abscesses. Every day, she had high, spiking fevers. And every day, she was treated around the clock with alternating doses of Tylenol and Motrin to suppress her fever.

Frustrated and concerned, the doctor taking care of this patient sat down with the parents, the child, and the nurses and explained that the continued treatment of fever could be blunting the child's attempt to rid her body of this harmful bacteria. Let's stop the antipyretics and give them only when the child feels that she just can't tolerate the fever. Everyone agreed. During the next two days, the nurses withheld antipyretics, and the bacteria disappeared from the bloodstream. Eventually, although it took weeks, the child recovered completely.

While it is certainly possible that the clearance of bacteria from the blood would have happened at the same time anyway, there was no convincing the parents or the child or the nurses of that. All believed that by withholding antipyretics, they were allowing the child's immune system to work better. And they were probably right.

MEDICATIONS that suppress fever are not necessary for our survival. On the contrary, medications that suppress fever only weaken our immune response. It's time to get over our fear of fever. Like Hippocrates, we would be much better served by embracing fever as an indispensable ally in our fight against infections instead of treating it like a harmful enemy.

2

Finishing the Antibiotic Course Is Often Unnecessary

We stopped looking for monsters under our bed when we realized that they were inside us.

—CHARLES DARWIN

In 1925, Sinclair Lewis published the book *Arrowsmith*, about a philandering, wisecracking, self-centered physician-scientist named Martin Arrowsmith, who discovers a virus that kills bacteria. Viruses that kill bacteria are called bacteriophages. (The suffix *-phage* is derived from the Greek word meaning "to eat or devour.") Arrowsmith uses a bacteriophage to end an outbreak of bubonic plague. The bacteriophage the author describes, however, isn't fictional. Lewis wrote this book about ten years before antibiotics were discovered. At the time, bacteriophages were often injected into patients suffering from severe or life-threatening illnesses, although it is unclear whether they actually worked.

Fast-forward to the present. On November 23, 2015, Steffanie Strathdee and her husband, Tom Patterson, went on a vacation to Cairo, Egypt. Both were epidemiologists at the University of California, San Diego, School of Medicine studying HIV/AIDS at the U.S.–Mexico border. A few days into their vacation, Tom got sick. It started with nausea, vomiting, stomach pains, high fever, a racing heartbeat, and dehydration. Patterson was taken by helicopter to Germany, where doctors found that he was infected with a bacterium called *Acinetobacter baumannii*. This particular strain—it was called Iraqibacter because more than three thousand wounded American and European soldiers fighting in the Middle East were diagnosed with it—was highly resistant to all commercially available antibiotics. When Patterson stabilized, he was airlifted back to the intensive care unit at the University of California, San Diego, Health Center, where he went into shock. During the next three months, while on a ventilator, Patterson lapsed in and out of consciousness. He lost more than one hundred pounds. Tom Patterson was dying.

Unclear what to do, Steffanie asked her husband if he was too tired to continue. "Just squeeze my hand if you want to live," she said to him. One minute passed. Then Patterson squeezed her hand, hard. Steffanie was determined to do what she could to save his life. As a first step, she went on the internet and found an article titled "Emerging Therapies for Multidrug-Resistant *Acinetobacter baumannii*." The article mentioned bacteriophages, the same therapy that had been described by Sinclair Lewis ninety years earlier. After penicillin was discovered, bacteriophage therapies disappeared. Now, however, in the face of growing antibiotic resistance, they were back.

Steffanie called Ryland Young, a researcher at Texas A&M, to see if bacteriophages could be given to her husband. As described in her book, *The Perfect Predator: A Scientist's Race to Save Her Husband from a Deadly Superbug*, Steffanie learned that bacteriophages weren't rare. About 10^{31}, or ten million trillion trillion, exist on the planet. And they're not hard to find. You just have to look in the right place—specifically, "sewage treatment plants, standing cesspools, or dumps filled with dirty diapers and other fecal-tainted debris, rotting garbage, the occasional dead animal, and wastewater runoff from a local hospital or animal farm."

Eventually, in coordination with teams from the U.S. Naval Medical Research Center; the Center for Phage Technology at Texas

A&M University; and AmpliPhi, a San Diego–based biotech company, a bacteriophage was found that killed Tom Patterson's strain of *Acinetobacter baumannii*. Although it was unclear how much of this bacteriophage to give, how often to give it, how it would be distributed, whether the bacteria would quickly become resistant, or whether Tom's body would make an immune response that killed the bacteriophage, the team decided to initially inject one billion bacteriophages into his bloodstream. First, however, they had to secure a compassionate-use exception from the Food and Drug Administration. Cara Fiore at the FDA facilitated the human experiment, sending Steffanie a legal document acknowledging that this therapy could precipitate her husband's death and that Steffanie would bear all responsibility.

In March 2016, four months after he had been injected again and again with one billion bacteriophage viruses, Tom Patterson emerged from his coma, his life saved. The good news is that the University of California, San Diego, among other groups, has now established a research center to design and test bacteriophages. The bad news is that the rise in antibiotic-resistant bacteria has made it a necessity.

Our distant past has predicted our future. The post-antibiotic era is now mimicking the pre-antibiotic era. "It's the microbes who will have the last word," wrote Louis Pasteur. Sadly, pharmaceutical companies are doing little to fight back. No new class of antibiotics has entered the market since 1980.

IMAGINE that you've just had knee surgery. Your doctor prescribes a pain medicine. After a few days, your pain is gone. Your doctor tells you that you can stop taking the medicine. You don't need it anymore.

Now imagine that you've had an asthma attack. Your doctor prescribes an inhaled medication to open up your breathing tubes. After a few doses, you can breathe again. Your doctor tells you that you can stop using the inhaler.

And now imagine that you've had fever, back pain, and cloudy urine. Your doctor takes a sample of your urine, tells you that you have a kidney infection, and starts you on an antibiotic. After a few days, your fever goes away, your urine becomes clear, and your back pain disappears. Still, your doctor insists that you finish the ten-day course of antibiotics. For the next seven days you continue taking the antibiotic even though you feel fine. Although it doesn't make much sense to you, according to your doctor, you still need it.

The dictum that it's important to complete a course of antibiotics is persistent and ingrained. In 2016, during "Antibiotic Awareness Week," the World Health Organization advised patients to "always complete the full prescription, even if you feel better, because stopping antibiotics early promotes the growth of drug-resistant bacteria." Similar advice has appeared in national campaigns in Australia, Canada, Europe, and the United States. In the United Kingdom, this advice is included in high school curricula.

In fact, the opposite is true. The longer the course of antibiotics, the *greater* the chance that bacteria will become resistant. Although the evidence couldn't be clearer, national campaigns to encourage people to "finish the antibiotic course" have been slow to catch up.

For most diseases, we stop taking medicines when we feel better. That's because all medicines have side effects; so, when symptoms are gone, the risks of the medicines outweigh their benefits. That is why doctors don't typically prescribe a "course" of pain medicines or a "course" of asthma inhalers. Why are antibiotics different? After all, the risks aren't insignificant. Antibiotics can cause severe allergic reactions, including hives, difficulty breathing, low blood pressure, and shock. One type of allergic reaction, called Stevens-Johnson syndrome, can be fatal. Prolonged use of antibiotics can allow for the overgrowth of a bacterium in the intestines called *Clostridioides difficile,* or C. *diff,* which can paralyze the colon (called toxic megacolon), an often fatal complication. Antibiotics increase the risk of yeast infections (e.g., thrush and vaginal infections) and the risk of kidney stones, especially in younger patients.

But the single greatest harm from prolonged and indiscriminate use of antibiotics is that it promotes bacterial resistance.

Still, we persist in giving antibiotics for days and sometimes even weeks longer than necessary. Why? It's not hard to imagine how this could have happened.

When antibiotics were first discovered, it was relatively easy to determine which antibiotics killed a particular strain of bacteria in the laboratory. This enabled doctors to choose the right antibiotic for an infection. It was also easy to figure out how long an effective amount of an antibiotic remained in the bloodstream. This enabled doctors to sort out how many times a day to give the antibiotic. But how were doctors supposed to determine how many days to give an antibiotic? One way would have been to have separated patients with the same bacterial infections into different groups; some would be treated for three days, others for five, seven, ten, or fourteen days. Given that patients treated for fewer days might fare poorly, or die, these studies were difficult to do. Instead, clinicians made their best guess, opting for the comfortably longer course. Now, however, because of the growing problem of antibiotic-resistant bacteria, we have been forced to reexamine our assumptions.

The misconception about how long to give antibiotics started with one of the first patients to receive them. In 1945, three men won the Nobel Prize for the discovery of penicillin: Alexander Fleming, Ernst Chain, and Howard Florey. Four years earlier, in 1941, Florey and Chain had treated their first patient, Albert Alexander, who had a small sore at the corner of his mouth that had spread to the rest of his face. The bacterium that had caused his infection was *Staphylococcus aureus* (staph). Within a few days, Alexander's entire face, scalp, and mouth were swollen so badly that doctors had to remove his eye to relieve the pain. Florey and Chain gave him 4 grams of penicillin—less than one day's worth of a modern dosing. To stretch the treatment course to four days, they recovered penicillin from Alexander's urine, purified it, and injected it back into his bloodstream. After the first day of treatment, Alexander's infection stabilized. After a few days, however, he worsened and died. The reason that Alexander succumbed to his infection wasn't that the staph had become resistant to penicillin or that four days of antibiotics weren't long enough; the problem was that the dose of penicillin was far too small to begin with. During the Nobel Prize ceremony, Alexander Fleming said, "If you use penicillin, use enough." Sadly, clinicians at the time interpreted this to mean that they should continue antibiotics until well after the patient appeared to be getting better, to avoid a relapse. The lengthy antibiotic course was born.

THE impact of antibiotics was dramatic and immediate. In 1900, the average life expectancy in the United States was forty-seven years. One hundred years later, the life expectancy had increased to seventy-four for males and eighty for females. Along with vaccines and purified drinking water, antibiotics have been largely responsible for this thirty-year increase.

Unfortunately, the golden age of antibiotics is drawing to a close. In the United States, about 2 million people are infected every year with bacteria that are highly resistant to antibiotics; 23,000 will die from these infections. Worldwide, the numbers are staggering. In April 2014, the WHO warned that "a post-antibiotic era—in which common infections and minor injuries can kill—is a very real possibility for the 21st century." The WHO now estimates that antibiotic-resistant bacteria will kill at least 50 million people per year by 2050.

In their quest to survive, bacteria have devised a variety of ways to resist antibiotics. They make antibiotic-destroying enzymes. They change proteins on their surface so that antibiotics can no longer attach. They create an internal pump that discharges antibiotics. Perhaps worst of all, these resistant bacteria teach neighboring bacteria to become resistant using small circular pieces of DNA called plasmids. This ability to exchange genetic information gets to the central horror of antibiotic resistance. It's often not the bacteria that doctors are treating that become resistant; it's other bacteria that live at the surface of the skin, nose, throat, and intestine that learn to become resistant. These bacteria (with names like *E. coli*, *Klebsiella*, and *Proteus*) are common causes of urinary tract, abdominal, and bloodstream infections.

Among the bacteria that have become resistant to antibiotics, perhaps none is more feared than methicillin-resistant *Staphylococcus aureus*, or MRSA. It is responsible for 19,000 deaths and 360,000 hospitalizations in the United States every year. In the past, MRSA was found almost exclusively in hospitals, where patients are given powerful antibiotics intravenously for long periods. Now, however, with the use of powerful antibiotics ingested by outpatients, the problem of MRSA has spread to the community.

The story of MRSA is particularly instructive. Before antibiotics, the mortality rate from severe staph infections was 80 percent. In the 1940s, when penicillin was first mass-produced, virtually 100 percent of the strains of staph were sensitive to it. By 1942, the first penicillin-resistant strains of staph appeared. Staph had learned to make an enzyme (called beta-lactamase) that destroyed the drug. Today, penicillin-sensitive strains of *Staphylococcus aureus* are rare. In the early 1960s, to combat the problem of penicillin resistance, scientists developed semisynthetic penicillins (e.g., methicillin), which weren't easily broken down by the enzyme. Staph changed its strategy, producing a gene (called *mecA*) that made it virtually impossible for methicillin to bind to and kill the bacteria. Within a few years, new strains of MRSA began to appear. Now we use an antibiotic called vancomycin to treat MRSA. Unfortunately, strains of MRSA that are completely resistant to vancomycin have also begun to appear. In 2014, Dr. Carl Nathan, a microbiologist at Weill Cornell Medicine, referring to the "golden era" of antibiotic development, wrote, "We call that period 'golden' because success seemed routine then; we call it an 'era' because it ended."

Apart from the fact that antibiotic-resistant bacteria are difficult to treat, they're also difficult to identify. And antibiotic-resistant bacteria often require antibiotics that are more expensive and more dangerous than those required to treat antibiotic-sensitive strains.

What follows is a heartbreaking example of just how bad the problem of resistant bacteria has become.

In the late 1990s, a teenage girl with severe cystic fibrosis was admitted to a hospital in the Northeast. Only a lung transplant could save her life. For months, she and her family had waited anxiously for a donor lung to become available. Then, miraculously, one arrived. The transplant was a success. Soon after the surgery, the child's lungs were functioning normally. Unfortunately, before the transplant, the child had been treated with many prolonged courses of antibiotics. As a consequence, she had become colonized with a strain of a bacterium (pseudomonas, common in patients with cystic fibrosis) that was resistant to all commercially available antibiotics. In an attempt to prevent this resistant strain from infecting her new lung, doctors had tried to eliminate it before the transplant by sponging her entire body with 5 percent acetic acid (vinegar). The attempts were in vain. Months after a successful lung transplant, the child succumbed to the resistant bacteria. The sight of a child being washed down with vinegar could easily have been a scene from a medical ward in the late 1800s. We have taken our first steps into a post-antibiotic era.

How can we preserve this precious resource? One way would be to stop giving antibiotics to patients who don't need them. Many antibiotic prescriptions are written for symptoms such as cough, runny nose, earache, and congestion—infections that are often caused by viruses for which antibiotics are useless. (Antibiotics treat only bacterial infections.) Recent estimates are that 30–50 percent of all antibiotic prescriptions are unnecessary. The actual number is probably closer to 75 percent. The all-too-easy-to-remember Z-Pak (azithromycin), currently the most prescribed antibiotic in the United States, is the prime offender. Remarkable, given that azithromycin is not a first-line antibiotic for any common bacterial infection.

Perhaps the best strategy to reduce antibiotic resistance is to abandon once and for all the archaic notion of the "antibiotic course" and give antibiotics only for as long as necessary. One of the first experts to publicly decry this problem was Dr. Martin Llewelyn, a professor of infectious diseases in the United Kingdom. In 2017, in an article written for the *British Medical Journal* titled "The Antibiotic Course Has Had Its Day," Llewelyn wrote that the length of treatments should be determined by how quickly the patient recovers, and should not be an arbitrary length determined at the beginning of the illness. No longer should we hold ourselves to the one-size-fits-all concept of the antibiotic course. The time had come, argued Llewelyn, to bring personalized medicine into the world of antibiotics.

Dr. Llewelyn's advice was built on a mountain of evidence for a variety of infections.

Treating Bacterial Infections in Adults

Appendicitis

Every year in the United States, about 300,000 people suffer from appendicitis. When the appendix bursts, bacteria spill into the abdomen, causing infections that are usually treated for ten to fourteen days. In 2008, Italian investigators studied ninety adults with appendicitis and found that three days of antibiotics were enough. In 2015, American investigators studied another five hundred adults and found that four days sufficed. Advisory groups now recommend only four or five days of antibiotics for a ruptured appendix. Nonetheless, in most instances, patients are treated longer.

Urinary tract infection

Urinary tract infections, which affect about 150 million people each year worldwide, can be divided into two groups: infections of the bladder (cystitis) and infections of the kidney (pyelonephritis). Typically these infections are treated with antibiotics for seven to fourteen days. In 2007, American investigators studied 340 adult women with cystitis, finding that three days of antibiotics were curative. In 2018, a multinational research team studied another 500 adult women with cystitis and found that five days were adequate. Advisory groups now recommend either three or five days of antibiotics for cystitis, depending on the antibiotic used. Again, however, many patients are treated longer than necessary.

Pyelonephritis can also be treated for far shorter lengths of time. In 2012, Swedish investigators studied 120 adult women with pyelonephritis and found that seven days of antibiotics were just as good as fourteen days. In 2013, Israeli investigators reviewed the results of eight trials involving another 2,500 adults who were treated for seven, ten, or fourteen days, finding no differences. In 2017, American investigators studied another 270 women and found that seven days were good enough. As a consequence of these studies, depending on which antibiotic is used, patients with severe pyelonephritis can now be treated for as short as five days and for no longer than seven days. Again, this recommendation is often ignored. It's hard to overcome the inertia resulting from our having used longer courses of antibiotics for decades.

Sinusitis

Sinusitis is another common inflammation for which adults are treated with antibiotics. In 2009, Greek investigators reviewed twelve studies involving 4,400 adults with sinusitis, finding no difference in outcomes for those treated for three or ten days. As a consequence, advisory bodies now recommend no more than five to seven days of antibiotics for adults with sinusitis. Of interest, in some people, sinusitis resolves without any antibiotics.

Pneumonia

Bacterial pneumonia continues to be a leading cause of hospitalization and death worldwide. In 2016, Spanish investigators performed a groundbreaking study. They divided three hundred adults with pneumonia into two groups: one group was treated until the patient's fever had returned to normal for forty-eight hours; the length of antibiotics for the second group was determined as the physician saw fit. The average length of antibiotics in the first group was five days and in the second group was ten days. Clinical outcomes were the same. *Stopping antibiotics when the patient started to feel better was sufficient.* Advisory committees now recommend only five days of antibiotics for patients with bacterial pneumonia, assuming they are feeling better and have normal vital signs, even for those who are hospitalized. Again, many doctors fail to follow this advice.

Skin infection

Like sinusitis, pneumonia, appendicitis, and urinary tract infections, skin infections such as cellulitis are incredibly common. The U.S. military alone observed more than a hundred thousand cases of cellulitis during a four-year period from 1998 to 2001. In 2004, investigators at the Brooke Army Medical Center, in Fort Sam Houston, Texas, treated ninety adults with cellulitis for either five or ten days. No difference. Advisory groups now recommend only five days of antibiotics for cellulitis.

Bloodstream infection (sepsis)

Even bloodstream infections, which if not treated properly can be fatal, may be treated for shorter lengths of time. In 2018, researchers from Israel published a study that divided six hundred patients with severe bloodstream infections into two groups. One group received the recommended fourteen days of antibiotics; the other, seven days. Investigators found no difference in outcomes.

ALL these studies had two things in common: all showed that we didn't need to treat bacterial infections for nearly as long as we had thought necessary, and all were performed in adults.

Treating Bacterial Infections in Children

But what about children? Although studies in children have not been nearly as extensive as those in adults, studies for two infections have been reassuring.

Urinary tract infection

In 2001, American investigators reviewed studies of 1,300 children with cystitis and found that three days of treatment sufficed. Similarly, in 2002, Australian investigators reviewed the results of ten studies of 650 children with cystitis who received antibiotics for either two to four or seven to fourteen days, again finding no difference in the time it took to eliminate bacteria from the urine and to resolve clinical symptoms. Advisory bodies now recommend two to four days of antibiotics for school-age children with cystitis without fever. Many pediatricians fail to follow this advice.

Ear infection

Ear infections are the single most frequently diagnosed bacterial infection in children in the United States, and the most common reason to receive antibiotics. In 2009, the WHO, in its review of existing studies, concluded that older children with ear infections can be treated successfully with antibiotics for three days.

TAKEN together, these studies show that for adults as well as for children old enough to describe their symptoms, the best advice for diseases such as cystitis, pyelonephritis, pneumonia, sinusitis, cellulitis, appendicitis, and ear infections is to stop antibiotics when symptoms begin to improve. When you think about what is happening in the body during these infections, this makes a lot of sense. Take pneumonia, for example. When bacteria such as pneumococci infect the lungs, they reproduce themselves. But it's not the bacteria that cause the symptoms of cough, fever, and difficulty breathing; it's the immune response to the bacteria—specifically, white blood cells flooding the lungs. The inflammation seen on chest X-rays is evidence of this response. When the immune system, in conjunction with antibiotics, has done its job, the signs and symptoms of inflammation, such as fever and difficulty breathing, abate. *At this point, bacterial replication is no longer part of the process*. Therefore, antibiotics are no longer necessary. The only thing that antibiotics are doing now is creating resistant strains and increasing the risk of side effects.

Another compelling reason to stop antibiotics that hadn't been appreciated until recently is that antibiotics alter something called the microbiome. Humans have about 100,000,000,000,000 (100 trillion) bacteria that live at the surface of their skin, intestines, nose, and throat. This colonization, the microbiome, contains an enormous quantity of bacterial genetic material. To the surprise of many (who had considered these bacteria to be harmless), researchers have now shown that certain microbiomes have been associated with conditions such as psoriasis, obesity, asthma, inflammatory bowel disease, colorectal cancer, reflux esophagitis, and heart disease.

Antibiotics can dramatically alter the microbiome, especially in children. Yet we manipulate the microbiome at our peril.

THE simple truth is that the length of antibiotic courses has always been somewhat arbitrary, usually involving multiples of five or seven—presumably because we have five fingers on each hand and there are seven days in a week. But bacteria don't replicate based on the metric system or the lunar month. Brad Spellberg, a physician working at the Keck School of Medicine at the University of Southern California, took the notion of a fixed one- or two-week course of antibiotics to its illogical end. He argued that in AD 321, the Roman emperor Constantine decreed that there would be seven days in a week. "Even in the modern era of evidence-based medicine," wrote Spellberg, "this 1,695-year-old decree remains a primary reference for the duration of antibiotic therapy." He referred to one or two weeks of antibiotic therapy as one or two "Constantine units."

All patients respond to their infections differently. Therefore, it makes sense to determine an antibiotic course based on the patient's response to the infection rather than on a preset length of time set at the beginning of the illness. Although it might feel wrong to treat bacterial pneumonia with 3.5 days of antibiotics, if the patient has had no fever for at least 2 days, that might be exactly the right thing to do. And it would conserve the only medicine we have to treat these severe and potentially fatal infections—a medicine that we are in the midst of losing.

IT is not unreasonable to assume that by 2050, when many commercially available antibiotics are rendered useless by resistant bacteria, medical wards might begin to look like those of one hundred years ago.

Walsh McDermott, a young physician working on the male medical ward at Bellevue Hospital in New York City in 1930, described a typical scene in the pre-antibiotic era:

> In the first five to ten beds would be the patient with pneumonia . . . He lies in bed frequently on the affected side. The eyes are bright, the face is flushed, and the earlobes, lips, and nail beds [are blue]. He [is] unable to [cough up] the [thick] bloody sputum, which has to be wiped from his lips. Save for the last few hours of life, he is usually conscious.
>
> Next to the patients with pneumonia would be one or two patients sitting up in their beds who had . . . heart failure from syphilis. Not infrequently, one might have a grapefruit-sized [lump] distorting the upper chest. This meant that . . . the aorta [had been destroyed and] might eventually burst.
>
> The next two beds would be occupied by patients in coma with meningitis. In their particular cases, the meningitis had arisen from

chronic ear infections. The course of meningitis was almost invariably fatal, and rapidly so.

In the next bed is a man with sepsis. A barber by occupation, he had repeatedly stuck his finger on a safety pin used to pin on his smock with each new customer. He would have essentially no hope of recovery.

Now, more than ever before, it is incumbent upon us to stop using antibiotics when we don't need them.

3

Antibiotic Drops Don't Treat Pinkeye

On June 17, 2013, CBS News reported an unusual outbreak of pinkeye among Japanese schoolchildren. The phenomenon came to national attention when a Japanese blogger interviewed a teacher who had observed two sixth-grade students licking each other's eyeballs in an equipment room. Confronted, they both admitted that eyeball licking—also known as "worming"—was popular in their class. In fact, it was apparently sweeping the nation.

Dr. Robert Noecker, an ophthalmologist practicing in Connecticut, explained the appeal. "The cornea is the most innervated part of the body," he said. "That's why it might feel good to have it licked. It's the same thing with sucking toes." Eyeball licking had become the new second base, when kissing gets boring.

Eyeball licking was, however, not without consequence. The Japanese teacher noted that at least ten students were wearing eye patches, presumably to hide their pinkeye. "This is a dangerous practice that has the potential to spread a number of bacteria that reside in the mouth," Robert Glatter, an emergency physician at Lenox Hill Hospital in New York, told CBS News. Glatter was particularly concerned about chlamydia, which, according to him, "had the potential to lead to blindness if untreated."

Bacterial infections, however, aren't the only cause of pinkeye. Eyeball licking could scratch the cornea, causing abrasions. Or pinkeye could have been caused by viruses or a chemical reaction to enzymes and acids in the mouth or an allergic reaction. No one knew.

It took two months to figure it out.

COVERING the eye is a thin, translucent membrane that begins at the edge of the cornea and extends to the inside of the upper and lower eyelids. The medical term for this lining is the *conjunctiva*. When the lining gets inflamed, people say they have pinkeye. Doctors call it conjunctivitis.

Conjunctivitis is not a trivial infection. Every year, about six million Americans suffer from this condition. About 70 percent of those who suffer will visit their internist or pediatrician or optometrist or ophthalmologist; many will go to emergency rooms or urgent care centers. Most will receive a prescription for antibiotic eye drops—at a cost to the health care system of between $377 and $857 million every year.

Conjunctivitis is so prevalent that states, schools, corporations, and day care centers have developed guidelines on how to treat it. Most of these institutions *require* people with conjunctivitis to receive antibiotic eye drops for at least twenty-four hours before returning to work or school. This requirement is designed to promote healing and prevent spread. As it turns out, it does neither. Here's why.

Conjunctivitis from Allergies

The most common cause of conjunctivitis is allergies. Every year, up to 40 percent of Americans suffer from allergic conjunctivitis.

Symptoms include itchy eyes; swollen eyelids; a thin, watery discharge; and a runny or itchy nose. Typically, the patient is allergic to cosmetics, contact lenses, pollen, dander, house mites, dust, pets, or other environmental allergens. Allergic conjunctivitis is most common during the spring and summer, when pollen counts are highest. And not surprisingly, it occurs in people who have other allergies, such as asthma or hay fever.

Because allergic conjunctivitis isn't a bacterial infection, antibiotics don't treat it. And because it isn't contagious, staying home from work or school doesn't lessen the chance of its spreading. The good news is that most people with allergic conjunctivitis understand what's happening to them. So, they're much less likely to go to a doctor and get a prescription for an antibiotic.

Conjunctivitis from Viruses

The second most common cause of conjunctivitis is viruses. In children, however, they're the most common cause. About 80 percent of infectious conjunctivitis is caused by viruses, most commonly one called adenovirus. Whereas allergic conjunctivitis usually occurs during the spring and summer, viral conjunctivitis occurs during the winter. Symptoms include a thin, watery discharge from the eyes and cold symptoms such as runny nose and sore throat. Unlike allergic conjunctivitis, viral conjunctivitis is highly contagious; between 10 and 50 percent of people who come in contact with someone infected will also be infected.

Because viral conjunctivitis, like allergic conjunctivitis, isn't a bacterial infection, antibiotics don't treat it. Also, because viral conjunctivitis usually lasts seven to ten days, staying home for only twenty-four hours before returning to work or school does little or nothing to lessen its spread.

Conjunctivitis from Bacteria

The third most common cause of conjunctivitis is bacteria. Whereas viral conjunctivitis causes a thin, watery discharge, bacterial conjunctivitis causes a thick, pus-filled (purulent) discharge that usually lasts all day, with eyes typically matted shut in the mornings. (A simple rule for distinguishing allergic, viral, and bacterial conjunctivitis is that allergic conjunctivitis causes itching, viral conjunctivitis is accompanied by other cold symptoms, and bacterial conjunctivitis causes a thick, purulent discharge. Unfortunately, there's a lot of overlap. So, it's often hard for doctors and patients to distinguish among them. As a consequence, many people with allergic or viral conjunctivitis receive antibiotics.)

Because bacterial conjunctivitis is caused by bacteria, it stands to reason that antibiotic eye drops should work. However, bacterial conjunctivitis is usually a self-limited disease, meaning that people recover from it whether or not they use antibiotic eye drops. So the real question is: Do antibiotic eye drops either shorten the duration of illness or lessen symptoms? If they do, then the requirement to treat bacterial conjunctivitis with antibiotic eye drops for twenty-four hours before returning to work or school makes sense. If they don't, then the requirement is unnecessary and, as we'll see, potentially harmful.

THE only way to determine whether antibiotic eye drops treat bacterial conjunctivitis is to do studies where one group receives antibiotic eye drops and the other receives drops that are antibiotic-free (placebo drops).

Antibiotics Don't Treat Bacterial Conjunctivitis

In 2012, researchers from the Centre for Population Health Sciences at the University of Edinburgh reviewed all the existing studies, writing, "We found 11 randomized controlled trials from different parts of the world which recruited a total of 3,673 participants overall. We judged two of the trials to be of high quality and we graded the remainder of poor quality." Of the two high-quality studies, one was performed in children, the other in adults.

The study performed in children was published in July 2005. Researchers from Oxfordshire in the United Kingdom divided 320 children with bacterial conjunctivitis into two groups. One group was given antibiotic eye drops every two hours for the first day and then four times a day until the infection had resolved. The other group was given placebo eye drops. Seven days later, the researchers found that 86 percent of the children in the antibiotic group and 83 percent in the placebo group were cured, an insignificant difference. The authors concluded, "Most children presenting with acute infective conjunctivitis in primary care will get better by themselves and do not need treatment with an antibiotic."

The study performed in adults was published later that year, in December 2005. Researchers from Switzerland and the Netherlands treated 160 adults with acute bacterial conjunctivitis with either antibiotic or placebo eye drops four times a day for seven days. At the end of the week, 62 percent of the treatment group and 59 percent of the placebo group were cured. Again, no significant difference. The authors concluded, "These findings do not support the current prescription practices of [antibiotic eye drops] by general practitioners."

Even the studies judged by the researchers at the University of Edinburgh to have been poorly done (some of which claimed that antibiotic eye drops worked) found differences that were far from dramatic. One study of adults showed a cure rate of 88 percent in the treatment group and 72 percent in the placebo group. Another in children found a cure rate of 91 percent in the treatment group and 72 percent in the placebo group. A third study, of adults and children, found a cure rate of 87 percent in the treatment group and 70 percent in the placebo group. And a fourth study, of adults and children, found a cure rate of 85 percent in the treatment group and 70 percent in the placebo group. In other words, most adults and children with acute bacterial conjunctivitis had completely resolved their symptoms within a week with or without antibiotic eye drops.

As a consequence of these studies, the American Academy of Ophthalmology now recommends watchful waiting for those suffering from acute conjunctivitis, whether it's caused by bacteria or not. Nonetheless, about 60 percent of people in the United States with conjunctivitis receive a prescription for antibiotic eye drops.

Because a handful of studies, even if poorly done, showed some benefit in the treatment of bacterial conjunctivitis, one could reasonably argue that the conservative choice would be to treat. After all, what's the harm of giving a few days of antibiotic eye drops? As it turns out, the decision to treat is more dangerous than most people realize.

- Antibiotic eye drops can be sensitizing, causing itching, burning, stinging, redness, and swelling—ironically, the same symptoms as conjunctivitis.
- Antibiotic eye drops can promote the emergence of resistant strains of bacteria. Because these resistant strains are now living on the surface of the body, they can cause more serious infections later, such as bloodstream infections or pneumonia, which, because they are now caused by resistant bacteria, are more difficult to treat.
- Antibiotic eye drops might also contain a steroid to decrease inflammation. About one out of every five people who receive antibiotic eye drops receives an antibiotic-steroid combination. Steroids, because they impair the immune response, will only prolong viral infections.
- Antibiotics change the types and quantities of bacteria that are always living on the surface of the eye. These bacteria play a protective role, making it more difficult for potentially harmful bacteria to gain a foothold. Antibiotics, however, make it easier.
- Antibiotic eye drops aren't free, costing about $60 per treatment course. For some of the newer antibiotics, the cost can be as high as $130.

IN the end, the requirement by many day care centers, schools, and workplaces to treat conjunctivitis with antibiotic eye drops for twenty-four hours won't treat bacterial infections, won't prevent the spread of viral infections, might be sensitizing, will be costly, might create antibiotic-resistant bacteria, and will only alter the normal bacteria living peacefully on the surface of your eye to protect against other bacteria that are far more dangerous. Patients are probably better off using lubricating eye drops or cold compresses to provide short-term relief of symptoms. Although this approach won't lessen the duration of the illness, at least it won't be harmful.

SO what about the Japanese eyeball-licking story? Did investigators ever figure out what specifically was causing the conjunctivitis?

In August 2013, two months after the story broke on CBS News, David Mikkelson from the fact-checking website Snopes took a closer look. Mikkelson found that the original story had appeared in a Japanese newspaper called *Bucchi News*. The article was titled "The Perverted Play of Eyeball Licking Is a Hit Among Primary Schoolers."

Bucchi News is produced by Core Magazine Co., a publishing company raided by police on suspicion of obscenity on April 19, 2013. A few days later, Core announced that it would suspend publication of two of its magazines. So, the source of the story wasn't particularly reliable. Also, no one had ever gone back to verify the truthfulness of the teacher who had observed the eyeball licking. Queries sent to organizations of Japanese ophthalmologists and school nurses revealed that none of them had ever heard of Japanese eyeball licking, a phenomenon that was supposedly sweeping the nation. Nonetheless, the story was picked up and reported as fact by CBS News, the *Guardian*, and *Medical Daily*, among others. In short, eyeball licking wasn't really a thing in Japan. Although it's interesting that there is actually a medical name for it: oculolinctus.

PART II
Supplements and Drugs

4

Vitamin D Supplements Aren't a Cure-all

In 2005, a forty-five-year-old woman, feeling fatigued, visited her doctor, who ordered a series of tests. She was told that the tests were all within the normal range, including her vitamin D level. Ten years later, the woman had another check of her vitamin D, again because she was feeling fatigued. The vitamin D level was unchanged. This time, however, the doctor told her that her level was inadequate, and recommended supplemental vitamin D. Same doctor, same testing laboratory, same vitamin D levels, same symptoms. What happened?

We're in the midst of a vitamin D craze. Everyone, it seems, is taking it. Vitamin D now outsells every other supplement on the market, including fish oil, probiotics, and multivitamins. Indeed, one in four older adults now takes vitamin D supplements.

The reason that so many people are taking vitamin D is to prevent thinning or weakening of their bones. But that's not all. They also take it to treat muscle weakness and fatigue or to boost their immune systems or to slow the aging process or to prevent cancer or heart disease or diabetes. Dr. Clifford Rosen, a bone specialist at Maine Medical Center, calls our compulsion to take vitamin D "a religion."

What most people don't realize is that they've been duped.

VITAMINS are essential for normal growth and nutrition. Unfortunately, our bodies can't make vitamins. We can get them only from food. Unlike every other vitamin, however, vitamin D *can* be made in the body. So, technically, it's not a vitamin; it's a hormone.

How do we make vitamin D? Sunlight converts a form of cholesterol in our skin to pre-vitamin D_3, which is then modified in the liver. When doctors request a blood test to measure vitamin D, it's this liver-modified form they're measuring, even though it's not the active form of the vitamin. The active form is made in the kidneys.

Vitamin D facilitates the absorption of calcium from the intestine. Without vitamin D, only 10 to 15 percent of dietary calcium is absorbed; with it, 30 to 40 percent is absorbed. Calcium is necessary to make bones strong; indeed, about 99 percent of the body's calcium is stored in bones. People who don't make enough vitamin D don't absorb enough calcium, which leads to bone thinning. This thinning or softening of the bones causes an outward bowing of the knees (rickets) in children and a dramatic increase in the incidence of fractures in older adults. About 50 percent of women and 25 percent of men over fifty will have a fracture due to bone thinning sometime during their life. So, it's not an insignificant problem.

Because vitamin D can be made in the body using ultraviolet light from the sun (hence its nickname the "sunshine vitamin"), we don't necessarily need to get vitamin D from food. This is fortunate because, unlike most other vitamins, it's hard to find a lot of natural vitamin D in what we eat. Vitamin D can be found in fatty fish such as salmon, mackerel, sardines, and tuna. It can also be found in beef liver, cheese, egg yolks, white beans, mushrooms, broccoli, collard greens, kale, and soy beverages. Because vitamin D isn't prevalent in nature, and because some people don't get adequate amounts of sunlight, years ago federal agencies required food producers to add vitamin D to milk, orange juice, yogurt, and breakfast cereals. Once foods were supplemented, the problem of severe vitamin D deficiency virtually disappeared in the United States. Rickets, which was common among children in the 1920s, has essentially vanished. Thinning of the bones in older adults, however, has not.

TO understand how we came to believe that so many people are now vitamin D deficient, we need to go back to the beginning.

Humankind's debut on earth occurred on a prehistoric landmass at or near the equator, where sunshine was plentiful. It was here that our ability to synthesize vitamin D in our skin was born. It wasn't long, however, before we migrated to areas where sunshine was less plentiful. Today, anyone who lives at or above 37 degrees north of the equator will have limited sunlight during the winter months. In the West, this translates to living north of a line extending from San Francisco to Philadelphia; in the East, from Athens to Beijing. To deal with this lack of sunlight, we developed a mechanism to store vitamin D in fat, which could then be liberated during the winter. This worked well for winters that were shorter than ten weeks. For longer winters, however, vitamin D stores were often depleted. At this point, vitamin D could be obtained only from food.

Given that most people get enough vitamin D from sunlight and that those who experience long winters get enough vitamin D from fortified foods, why does anyone need to take vitamin D supplements? The answer can be found in the influence of two different types of studies, neither of which, as it turns out, was correct.

First, researchers compared the relative health of people who lived at or near the equator with those who lived farther away. Studies that compare health outcomes in people who live in different parts of the world are called ecological studies. Researchers found that those who lived farther from the equator were more likely to suffer from asthma, diabetes, multiple sclerosis, obesity, schizophrenia, bipolar disorder, and seasonal affective disorder (SAD); they were also more likely to develop colon, breast, and prostate cancer. Most impressively, a twenty-year study of thirty thousand people found that those with limited sun exposure didn't live as long. Vitamin D, it appeared, was far more important to human health than simply making strong bones.

Then researchers took the next step. To prove that these dramatic differences in health were *caused* by vitamin D, they compared the health of people who took vitamin D supplements with those who did not. This time, however, it didn't matter where people lived or how much they went out in the sun. Studies that compare the health of one group with another where those being studied (not the investigator) decide whether to take a particular medicine or dietary supplement are called observational studies. In this case, the investigator stands back and just observes what is happening.

Again, the results were clear. Observational studies showed that those who took vitamin D supplements were less likely to develop breast, ovarian, skin, prostate, lung, and colon cancer. They were also less likely to have high blood pressure, more likely to have high levels of high-density lipoprotein cholesterol (HDL, the good cholesterol), and less likely to suffer heart attacks. Further, they were less likely to be stricken by infectious diseases such as tuberculosis and influenza or to be diagnosed with mental disorders such as depression, Alzheimer's disease, and Parkinson's disease. They were also less likely to suffer from diseases such as diabetes, multiple sclerosis, rheumatoid arthritis, and thyroid disease as well as a variety of other problems such as obesity, chronic kidney diseases, age-related macular degeneration of the eye, chronic pain syndromes, and fractures from falls. In other words, people who took supplemental vitamin D were less likely to be burdened by almost every disease known to humankind.

How was this possible? If vitamin D affects only the amount of calcium in bones, why would it impact so many other systems in the body? A closer look at the biology of vitamin D appeared to provide the answer. Scientists found that most cells in the body have their own receptors for vitamin D and that when vitamin D binds to cells, it can impact more than two hundred different genes that regulate cellular activity. Further, they found that the enzyme that converted vitamin D to its active form wasn't found only in the kidneys; it was found also in the colon, prostate, and breast, which might explain why vitamin D lowered the incidence of cancer at those sites. Finally, researchers found that in addition to its effect on bone metabolism, vitamin D impacted the immune system and other hormonal systems as well as how cells grew and differentiated.

Everything added up. The only thing left was to prove that vitamin D was doing exactly what the ecological and observational studies had shown that it was doing. Researchers then turned to a third group of studies called randomized, controlled clinical trials. Unlike the ecological studies, which observed health differences based on where people lived, or observational studies, which allowed subjects to choose whether they took vitamin D or not, researchers performing randomized, controlled trials determined who would receive supplemental vitamin D. Investigators could now make sure that those who did or did not take vitamin D supplements were identical in their medical backgrounds, socioeconomic status, health care–seeking behavior, and areas of residence.

Again, as was true for the ecological and the observational studies, the results from the randomized, controlled clinical trials were clear. This time, however, they were exactly the opposite of what researchers had expected.

Vitamin D Doesn't Treat Cancer

In 2019, researchers from Tokyo divided four hundred patients with cancers of the esophagus, stomach, colon, or rectum into two groups. One group received 2,000 international units (IU) of vitamin D every day; the other, a placebo. Researchers found no differences in the survival rates or relapse rates for any cancer.

Vitamin D Doesn't Prevent Cancer

In 2014, researchers from Denmark reviewed the results of eighteen studies involving 50,000 middle-aged women. The cancer rate was 7.6 percent in those who took vitamin D and 7.7 percent in those who didn't. The only difference between these two groups was that those who took vitamin D had an increased risk of kidney stones.

In 2017, researchers from Creighton University in Nebraska reported the results of a four-year study of vitamin D involving 2,300 middle-aged women who did or didn't receive vitamin D plus calcium. The incidence of cancer in the two groups was indistinguishable.

In 2018, researchers from the United Kingdom reviewed the results of thirty studies of vitamin D involving 19,000 men and women followed for up to six years. The incidence of cancer and cancer-related deaths was the same in both groups.

In 2019, researchers from Harvard Medical School reported the results of a five-year study in which they divided 26,000 older adults into two groups. One group received 2,000 IU of vitamin D every day; the other group didn't. At the end of the study, the incidence of cancer was the same in both groups.

Vitamin D Doesn't Prevent Stroke or Heart Disease

In 2014, researchers from New Zealand reviewed the results of eight studies involving 46,000 participants, finding that vitamin D with or without calcium was of no value in preventing stroke or heart disease. They also reviewed the results of nine studies with a total of 49,000 participants and found no benefit in the prevention of heart attacks.

Vitamin D Doesn't Prevent Coughs, Colds, or Sore Throat

In 2012, researchers from New Zealand studied 320 healthy adults who either did or did not receive vitamin D monthly for eighteen months. The incidence of respiratory infection was the same in both groups. No differences were seen in the number of infections, the severity of symptoms, the duration of symptoms, or work days missed.

Vitamin D Doesn't Prevent a Variety of Other Diseases

In 2017, French researchers reviewed eighty-three studies, finding no differences in the incidence of obesity, diabetes, mood disorders, muscular function, tuberculosis, or cancer in those who did or didn't take vitamin D with or without calcium.

Vitamin D Doesn't Prevent Fractures

In 2010, Australian researchers studied 2,300 women over age seventy who either were or were not given vitamin D every autumn for three to five years. Shockingly, those who received vitamin D were *more* likely to fall and *more* likely to suffer fractures than those who didn't.

In 2014, French researchers reviewed the results of twenty-two studies involving a total of 76,000 participants, finding no differences in the incidence of fractures in those who did or didn't receive vitamin D with or without calcium. They also reviewed the results from twelve trials involving 28,000 participants who did or didn't receive vitamin D, again finding no differences in the incidence of hip fracture.

In 2017, researchers in China reviewed the results of thirty-three studies involving 51,000 participants over age fifty who either did or didn't receive vitamin D plus calcium. They found no differences in the incidence of fractures regardless of the gender, history of previous fractures, dietary calcium intake, baseline vitamin D levels, or dose of vitamin D and calcium given.

In 2018, researchers from New Zealand reviewed eighty-one studies involving 54,000 people who did or did not receive vitamin D supplements. They found that vitamin D had no effect on the number of total fractures, hip fractures, or falls, or on bone mineral density.

Worse, researchers performing randomized, controlled studies found not only that vitamin D didn't work but that high blood levels of vitamin D could cause weakness (which might explain why the 2010 Australian study found an *increased* risk of falls resulting in fractures), nausea, vomiting, frequent urination, and kidney stones.

What happened?

Why the Study Results Conflict with One Another

Observational studies didn't account for critical differences between people who did or didn't take vitamin D

In observational studies, individuals decide whether to take vitamin D. In randomized, controlled studies, researchers make that decision. This is an important difference. As it turned out, people who chose to take vitamin D were more likely to be wealthier, to have health insurance, to visit a doctor when sick, and to exercise; they were also less likely to smoke.

Vitamin D is a marker for poor health, not a cause of poor health

When one factor (e.g., vitamin D levels) is associated with a particular disease but not the *cause* of that disease, it's called an epiphenomenon. One notable example of an epiphenomenon is something called the anemia of chronic disease. Many people with inflammatory diseases (e.g., infections, rheumatologic diseases, or cancer) are anemic, meaning that they have low numbers of circulating red blood cells. In these cases, because the anemia isn't causing the disease, giving blood transfusions won't treat it. Similarly, if low levels of vitamin D are just a marker for chronic illness, it shouldn't be surprising that giving vitamin D won't treat or prevent that illness.

Vitamin D levels alone cannot determine whether someone is vitamin D deficient

In the United States, no group has been diagnosed more frequently or more inaccurately as vitamin D deficient than African Americans. Because vitamin D levels are consistently lower in black people than in white people, black people are often told that they need to take supplemental vitamin D. At the same time, however, their bone mineral density is significantly *higher* than that in white subjects. How is this possible? In a paper published in the prestigious *New England Journal of Medicine,* Harvard researchers found the answer. Although black people clearly have lower levels of vitamin D, they also have much lower levels of vitamin D–binding protein, something that is used by the body to store vitamin D in the winter. Because less vitamin D is bound by this protein, more is available to absorb calcium from the intestine and build strong bones. By choosing to measure only vitamin D levels in black patients, doctors have consistently misdiagnosed those patients.

The failure of physicians to measure vitamin D–binding protein is only part of the problem. In addition to its importance in building strong bones, calcium is also critical to the normal function of electrical impulses in the heart. Too much or too little calcium in the bloodstream could cause the heart to stop beating. Because specific levels of calcium in the bloodstream are essential to life, it's a highly regulated process. And neither vitamin D nor vitamin D–binding protein is doing the regulating. Rather, both are regulated by a gland located behind the thyroid, in the neck. This gland, called the parathyroid gland, releases a substance called parathormone (PTH), which regulates the conversion of vitamin D to its active form in the kidney. If calcium levels in the bloodstream are too low, PTH liberates calcium from bones, even if that means that the bones become dangerously thin. The body would rather suffer weakened bones than a heart that doesn't beat properly.

The interplay among the synthesis of vitamin D by the sun, liver, and kidneys; the storage of vitamin D in fat by vitamin D–binding protein; and the regulation by PTH of both vitamin D and calcium is complex. So, when doctors get a vitamin D level, they are looking at only one part of the process, which can be misleading. It would be far more accurate to measure calcium, PTH, vitamin D–binding protein, and vitamin D at the same time. It is the rare doctor who does this.

Vitamin D isn't the only chemical enhanced by sunlight

Studies showing that people who lived at or near the equator were healthier than those who lived farther away were misleading for another reason. In addition to vitamin D, sunlight also enhances the production of a variety of other substances, such as beta-endorphins, which modify pain; calcitonin, which works in opposition to PTH; substance P, which affects blood flow, inflammation, pain, mood, anxiety, and cell growth; adrenocorticotropic hormone (ACTH), which controls the immune system and inflammation; and melanocyte-stimulating hormone, which reduces appetite, increases libido, and is responsible for skin pigmentation. All these substances affect health, all are enhanced by exposure to the sun, and none was measured in studies that determined the effect of sun exposure on human health.

WITH all this information in hand, two federal advisory bodies, the Institute of Medicine (IOM) and the U.S. Preventive Services Task Force (USPSTF), convened a panel of experts to determine who should be tested for vitamin D deficiency and what levels of vitamin D in the bloodstream were adequate. Both groups reached the same conclusions: (1) Exposure to sunlight and fortified foods alone provide enough vitamin D for 97.5 percent of the U.S. population; (2) blood levels above 20 nanograms (a nanogram is one billionth of a gram) of vitamin D are adequate; and (3) the only people who should be tested for vitamin D deficiency are those with thinning bones; those with conditions that affect the absorption of fat from the intestine, such as weight-loss surgery or celiac disease; and those taking medicines that affect the absorption or processing of vitamin D such as steroids and antiseizure medications. For people unable to absorb vitamin D from food, the IOM recommended the daily intake of 600 IU up to seventy years of age and 800 IU for those older. *Everyone else should stop getting tested for vitamin D and stop taking vitamin D supplements.*

So, if, according to the IOM and the USPSTF, most people have adequate levels of vitamin D and get enough vitamin D from the sun and from fortified foods, why are so many people now told that they are vitamin D deficient? The reason is that although the IOM and the USPSTF agreed that levels of vitamin D greater than 20 nanograms were adequate, most laboratories (e.g., Quest Diagnostics and LabCorp) list 20 to 30 nanograms as *insufficient*. Why?

On April 10, 2017, Gina Kolata pulled back the curtain. In an article written for the *New York Times* titled "Why Are So Many People Popping Vitamin D?" Kolata noted that Dr. Michael Holick, a professor of medicine, physiology, and biophysics at Boston University School of Medicine, was a "leading proponent of the idea that just about everyone needs a vitamin D supplement." Contrary to the IOM and the USPSTF, Holick believed that people needed vitamin D levels to be at least 30 nanograms, not 20. This level cannot easily be achieved by diet and would require almost constant exposure to the sun. The only way that these higher levels could be obtained would be by taking vitamin D supplements. It's not unusual to find the occasional doctor who disagrees with recommended federal guidelines. But Michael Holick wasn't just any doctor. In 2011, Holick headed a committee of the Endocrine Society that recommended that vitamin D levels be at least 30 nanograms, which meant that about half the U.S. population was suddenly vitamin D deficient. As a consequence of this recommendation, commercial labs ignored federal advisory groups and began listing levels of 20 to 30 nanograms of vitamin D as insufficient. Most continue to do so today.

Holick's advocacy has been embraced by Gwyneth Paltrow's lifestyle and wellness company, Goop, and by Mehmet Oz, who describes vitamin D as "The No. 1 thing you need more of." Oz tells his audience that vitamin D will help avoid heart disease, depression, weight gain, memory loss, and cancer, despite abundant evidence to the contrary.

Ravinder Singh, who runs a testing lab at the Mayo Clinic, was shocked by the sudden change in recommended levels. "Demand for vitamin D testing went through the sky," he said. "It was almost as though there was nothing else in clinical practice."

On August 18, 2018, Liz Szabo, in an article for the *New York Times* titled "Vitamin D, the Sunshine Supplement, Has Shadowy Money Behind It," took a closer look at exactly who was funding Dr. Michael Holick. Szabo noted that Holick's guidelines had helped push vitamin D sales up to $936 million a year, a ninefold increase over the previous decade. Holick acknowledged in an interview that he was receiving $1,000 a month from Quest Diagnostics, a commercial lab that was clearly benefiting from his recommendations. Holick also had extensive ties to the pharmaceutical industry. Between 2013 and 2017, he received $163,000 for consulting services, according to Medicare's Open Payments Database, which tracks payments from drug manufacturers. The companies that paid him include Sanofi-Aventis, which markets vitamin D supplements; Shire, which makes drugs for hormonal disorders that are treated with vitamin D; Amgen, which makes a drug for the treatment of bone thinning; and Roche Diagnostics and Quidel Corporation, both of which make vitamin D tests.

Dr. Holick's ties to the tanning bed industry have also come under fire. In 2008, David Armstrong, in an article for the *Wall Street Journal* titled "Researcher Received Industry Funds," noted that

Holick had published an article recommending the moderate use of tanning beds to increase vitamin D levels—no doubt to the dismay of dermatologists, who know that tanning beds increase the risk of skin cancer. According to the *Wall Street Journal* article, Boston University had received a series of grants totaling $162,014 from the UV Foundation, an organization funded by the now-defunct Indoor Tanning Association, whose board of directors was composed entirely of tanning bed industry executives, in order to fund vitamin D research by Holick. For this reason, Holick was asked to step down from his position in Boston University's dermatology division, which he did. But he remained a professor in the medical school's Department of Endocrinology, Diabetes, Nutrition, and Weight Management. Although Holick argued that his industry funding "doesn't influence me in terms of talking about the health benefits of vitamin D," his financial association with a testing laboratory that has directly benefited from his advocacy is, at best, unseemly.

The clash between the levels of vitamin D listed by testing laboratories and those recommended by the two federal advisory panels isn't the only force working against patients. Advocacy organizations such as the U.S. National Osteoporosis Foundation (NOF) and the Europe-based International Osteoporosis Foundation (IOF)—*osteoporosis* is the medical term for thinning bones—also ignore federal guidelines and constantly promote the notion that only vitamin D levels above 30 nanograms are adequate and that otherwise healthy people should be routinely screened. Twelve of the twenty-two NOF corporate sponsors and fourteen of the twenty-five IOF corporate sponsors sell nutritional products that promote vitamin D.

Consumers today have a choice to make. Either they can believe that both the IOM and the USPSTF are correct and that 97.5 percent of the U.S. population are getting adequate amounts of vitamin D from sunlight and fortified foods; or they can believe that diagnostic labs are correct and that roughly half the U.S. population (about 165 million people) is vitamin D deficient. Dr. JoAnn E. Manson, a preventive medicine researcher at Brigham and Women's

Hospital in Boston, summed it up best: "A lot of clinicians are acting like there is a pandemic," she said. "This gives them justification to screen everyone." In the end, patients should ask themselves who is really benefiting from all this testing, all these prescriptions, and all these office visits.

5

Supplemental Antioxidants Increase the Risk of Cancer and Heart Disease

I don't want to achieve immortality through my work. I want to achieve it through not dying.

—WOODY ALLEN

We are born to die. And although there are many reasons for this, one is inescapable: oxidation.

All cells contain small organelles called mitochondria, where food is converted to energy. This conversion, which is necessary for cells to function, requires oxygen (oxidation). Oxidation invariably generates something called free radicals, electron scavengers that damage DNA, cell membranes, and the lining of arteries. Not surprisingly, free radicals cause cancer, aging, and heart disease. Although free radicals can be found in a variety of environmental sources (e.g., cigarette smoke, heavy metals, and toxins), the one source that is inescapable is our own mitochondria. To fight free radicals, our bodies make antioxidants (such as glutathione). But that's not enough. We also benefit from supplementing our own antioxidants with antioxidants contained in food (such as vitamins A, C, E, lycopene, and beta-carotene).

AS you might expect, people who eat more fruits and vegetables containing antioxidants have lower rates of cancer and live longer. This isn't controversial. Scientists have known about it for more than fifty years.

Antioxidants in Foods Reduce the Risk of Cancer

Between 1967 and 1973, researchers from Oslo, Norway, studied the dietary habits of 8,000 men. Those who ate foods containing vitamin A had a lower risk of lung cancer than those who didn't. Two subsequent studies performed in the United States, one in 2,000 middle-aged men, the other in 17,000 men and women, found the same thing. In 1989, researchers at the National Cancer Institute in Bethesda offered further proof. They found that people with higher levels of beta-carotene (a vitamin A precursor) in their bloodstreams had a lower risk of lung cancer.

By 1996, more than two hundred human studies and twenty-two animal studies had shown that diets rich in fruits and vegetables protected against a variety of cancers. After reviewing all these studies, researchers from the World Cancer Research Fund in London concluded, "The evidence for a protective effect of greater vegetable and fruit consumption is consistent for cancers of the stomach, esophagus, lung, oral cavity [mouth], pharynx [throat], endometrium [uterus], pancreas, and colon." By the late 1990s, the subject had been so well studied that researchers could pinpoint exactly which fruits and vegetables were the most beneficial. In descending order of importance, the list included raw vegetables, allium vegetables (onions, garlic, leeks, chives, and shallots), carrots, green vegetables, cruciferous vegetables (cauliflower, cabbage, broccoli, Brussels sprouts, garden cress, and bok choy), and tomatoes.

THE stage was set. If antioxidants in foods prevented cancer, it stood to reason that supplemental antioxidants, which contained far greater quantities than could ever be found in foods, would have an even greater protective effect.

Unfortunately, it hasn't worked out that way.

Supplemental Antioxidants Don't Lower the Risk of Cancer and Heart Disease

In 1996, researchers from Harvard Medical School divided 22,000 male physicians into two groups. One group received 50 milligrams of beta-carotene every other day for twelve years; the other, a placebo. No differences were found between the two groups in the incidence of lung cancer, the number of deaths from any type of cancer, the number of deaths from heart disease, the number of deaths from any cause, or the incidence of heart attacks or strokes.

In 1999, researchers from Helsinki, Finland, divided 29,000 male smokers into four groups. One group received vitamin E, 50 milligrams a day; a second group received beta-carotene, 20 milligrams a day; a third group received both these supplements; and a fourth group received neither. At the end of eight years, the researchers found no differences in the incidence of cancers of the pancreas, kidney, bladder, colon, rectum, mouth, throat, esophagus, or windpipe. Perhaps most surprising, *men who consumed supplemental beta-carotene had an 18 percent greater risk of lung cancer than those who had received a placebo.* The researchers were unsure of what to make of this finding. "Our results raise the possibility that supplemental beta-carotene may be harmful to smokers," they admitted. In the end, they were convinced that they had witnessed a meaningless statistical aberration. "This finding may well be due to chance," they wrote.

In 1999, researchers from Brigham and Women's Hospital and Harvard Medical School divided 40,000 women into two groups. One group received 50 milligrams of beta-carotene every other day; the other, a placebo. After four years, the researchers found no differences between the two groups in the incidence of cancer, heart disease, or overall mortality. Later, using the same population of women, researchers found that vitamin E, 600 international units (IU), given every other day for ten years, also didn't impact the incidence of cancer or heart disease.

In 2000, researchers from Ontario, Canada, divided 10,000 men and women age fifty-five or older who had heart disease or diabetes into two groups. One group received vitamin E, 400 IU, every day; the other, a placebo. Seven years later, the incidence of cancer, cancer deaths, strokes, heart attacks, and deaths from heart disease was the same in both groups. Supplemental antioxidants hadn't helped those at highest risk from strokes and heart attacks.

During the next few years, researchers across the globe, in an attempt to find some evidence that dietary supplements prevented cancer and heart disease, redoubled their efforts. Surely, if studies were done the right way, with the right combination of supplements at the right dose, a protective effect could be found. Again, however, they failed.

By 2003, fourteen more studies involving 170,000 people given combinations of vitamins A, C, E, and the mineral selenium or a placebo had been performed. Reviewing these studies, many of which involved taking maximal doses of supplements for long periods of time, researchers from Copenhagen, Denmark, found no evidence that antioxidants prevented cancers of the esophagus, stomach, colon, rectum, pancreas, or liver.

In 2004, researchers in Paris divided 13,000 men and women into two groups. One group received a capsule containing 120 milligrams of vitamin C, 30 milligrams of vitamin E, 6 milligrams of beta-carotene, 100 micrograms of selenium, and 20 milligrams of zinc; the other, a placebo. Seven years later, the researchers found no differences between the two groups in the incidence of cancer, heart disease, or death from any cause.

In 2009, researchers from Brigham and Women's Hospital and Harvard Medical School divided 15,000 male physicians into two groups. One group received vitamin E, 400 IU, plus vitamin C, 500 milligrams, every day; the other, a placebo. At the end of eight years, the researchers found no differences between the two groups in the incidence of prostate, lung, colon, and rectal cancers.

In 2013, the U.S. Preventive Services Task Force, in a masterstroke of understatement, concluded, "Limited evidence supports any benefit from vitamin and mineral supplementation for the prevention of cancer and [heart] disease."

WHAT happened? Why did foods containing antioxidants clearly prevent cancer while supplemental antioxidants did not? Scientists have offered two explanations.

First, milieu matters. Antioxidants in food and supplemental antioxidants in tablets aren't the same thing. Vitamins are phytochemicals, which means that they are contained in plants. (*Phyto* is the Greek word for "plant.") Vitamins in food are surrounded by thousands of other phytochemicals. Vitamins synthesized by companies and compressed into a tablet are not. It's likely that these other chemicals in food are also important. For example, as described in Catherine Price's book *Vitamania*, half an apple has the antioxidant activity of 1,500 milligrams of vitamin C, even though it contains only 5.7 milligrams of the vitamin. What this means is that if you choose to eat an apple, which is a product of nature, the impact of the antioxidants in it is far greater than if you ingest a pill, which is a product of the pharmaceutical industry. Nature is better.

Second, it's not the antioxidants in food that are preventing cancer; it's something else. Barry Halliwell, a professor of biochemistry at the National University of Singapore, in an article titled "The Antioxidant Paradox," wrote, "A protective effect of diet is not equivalent to a protective effect of antioxidants in diet." Although fruits and vegetables are rich in antioxidants, these foods contain other phytochemicals, all of which might be having an effect. Specifically, flavonoids, flavanols, flavanones, isoflavones, anthocyanins, anthocyanidins, proanthocyanidins, tannins, isothiocyanates, carotenoids, allyl sulfides, polyphenols, phenolic acids, indole-3-carbinol, allium compounds, protease inhibitors, saponins, phytosterols, inositol hexaphosphate, D-limonene, lutein, folic acid, lycopene, and dietary fiber—to name a few.

ALTHOUGH supplemental antioxidants don't appear to protect against cancer and heart disease, what's the harm? At worst, supplements make for a lot of expensive urine, but are they hurting people? Yes, they probably are. Five large studies performed during the past twenty-five years have shown that not only do supplemental antioxidants fail to prevent cancer and heart disease, they cause it. The Finnish researchers who had dismissed their observation that beta-carotene increased the risk of lung cancer in male smokers had unknowingly provided an early insight into what was really happening.

Supplemental Antioxidants Increase the Risk of Cancer and Heart Disease

In 1996, researchers from the Fred Hutchinson Cancer Research Center in Seattle, Washington, divided 18,000 smokers, former smokers, and asbestos workers into two groups. One group received beta-carotene, 30 milligrams, plus vitamin A, 25,000 IU, every day; the other, a placebo. The researchers reasoned that because this population was at the greatest risk for lung cancer, they would be the ones most likely to benefit. Much to their surprise, they found that the antioxidant group had a 28 percent increased risk of lung cancer, a 46 percent increased risk of dying from lung cancer, a 26 percent increased risk of death from heart disease, and a 17 percent increased risk of death from any cause compared with the group that had received a placebo capsule! For this reason, the study was discontinued twenty-one months earlier than planned. "The results of the trial were troubling," the researchers concluded. Six years after the supplements had been discontinued, the increased risk of lung cancer and death from all causes persisted. Eleven years later, researchers found that the supplement group also had a 52 percent increased risk of aggressive prostate cancer.

The finding that vitamin A and beta-carotene dramatically increased the risk of lung cancer, prostate cancer, and heart disease worried researchers. Could this also be happening with other vitamin supplements? Quickly, they turned their attention to vitamin E.

In 2005, researchers from Johns Hopkins School of Medicine in Baltimore, Maryland, reviewed nineteen trials involving 140,000 subjects who had received either supplemental vitamin E alone or in combination with other vitamins and minerals. They found that the supplement groups had a greater incidence of death from any cause. Further, the higher the dose of vitamin E, the greater the risk. The authors concluded that "High-dosage vitamin supplementation is often assumed to be at worst innocuous. In view of the increased mortality associated with high dosages of beta-carotene, and now vitamin E, *use of any high-dosage vitamin supplements should be discouraged until evidence of efficacy is documented*" (emphasis mine).

Then the Hopkins researchers offered one more piece of advice: "Policymaking bodies . . . should also caution the public against the use of high-dosage vitamin E supplementation." Their warning, however, has been largely ignored.

The same year that the Hopkins researchers released their findings, researchers from Laval University Cancer Research Center in Quebec divided five thousand men into two groups. One group received supplemental vitamin C, vitamin E, beta-carotene, selenium, and zinc; the other, a placebo. Subjects were followed for eight years. The hope was that supplemental antioxidants would decrease the risk of prostate cancer in men at highest risk, meaning those with elevated levels of prostate-specific antigen (PSA). They didn't. Again, supplemental antioxidants actually increased the risk.

In 2007, French researchers divided 13,000 men and women into two groups. Mimicking the Quebec study, one group received vitamin C, vitamin E, beta-carotene, selenium, and zinc; the other, a placebo. This time, however, researchers didn't find an increased risk of prostate cancer in men; they found an increased risk of skin cancer, specifically melanomas, in women. The authors concluded, "It clearly appears more cautious to recommend a lifelong healthy diet including a high consumption of fruits and vegetables able to provide the antioxidants required, rather than uncontrolled antioxidant supplement use."

In 2009, researchers from the Cleveland Clinic and several other U.S. cancer centers divided 36,000 men from the United States, Canada, and Puerto Rico into four groups. All were at least fifty years old, and all had normal PSA levels, meaning they were presumably at low risk of developing prostate cancer. One group received vitamin E, 400 IU a day; a second group received selenium, 200 micrograms a day; a third group received both supplements; and a fourth group, neither. At the end of the study, 620 men in the supplement group and 529 men in the placebo group developed prostate cancer. Again, supplements had increased the risk. The authors, noting that more than 50 percent of American men were taking supplemental vitamin E, warned, "The observed 17 percent increase in prostate cancer incidence demonstrates the potential for seemingly innocuous yet biologically active substances such as vitamins to cause harm. [This] underscores the need for consumers to be skeptical of health claims for unregulated over-the-counter products."

THE finding that antioxidant supplements actually increased the risk of cancer shouldn't have been surprising. Three studies of multivitamins performed in the 1980s and '90s, which had managed to fly below the radar, had found the same thing.

Between 1982 and 1989, the Centers for Disease Control studied more than a million Americans and found that male smokers who took multivitamins had an increased risk of cancer. Similarly, a study of 475,000 men performed by the American Cancer Society found that those who took multivitamins had a 15 percent increased risk of dying from prostate cancer. Finally, a study performed by the National Cancer Institute of 300,000 men found that those who took multivitamins more than seven times a week also had an increased risk of prostate cancer. All these studies were observational. Patients, not researchers, decided whether they were going to take supplements and which ones they were going to take. Some of the researchers involved in these studies argued that people who were sicker were more likely to take supplements. Although the increased risk of cancers found in the multivitamin studies was worrisome, the results were largely ignored.

Studies of experimental animals had also been predictive. Three studies in mice, one performed at the University of Texas Southwestern Medical Center in Dallas and the others at the University of Gothenburg in Sweden, found that supplemental antioxidants actually increased the rate of progression of melanomas and lung cancer.

STILL, the question remained. Why? Why do supplemental antioxidants increase the risk of heart disease as well as so many different types of cancer? Scientists have postulated the following series of events.

Vitamins are essential to health. Without them, people suffer diseases such as scurvy (a deficiency of vitamin C) or pellagra (a deficiency of vitamin B_3 [niacin]). For this reason, the National Academy of Medicine determines the Recommended Dietary Allowance (RDA) for a variety of vitamins and minerals. Researchers studying supplemental antioxidants, however, gave doses far greater than the RDAs, far greater than those found in nature, and all, unfortunately, far too available over the counter. For example, the RDA for vitamin C is 60 milligrams. But most people who take vitamin C supplements ingest 1,000 milligrams—more than sixteen times the RDA. A person would have to eat eight cantaloupes or fourteen oranges to get the same amount of vitamin C contained in a single 1,000-milligram tablet. For vitamin E, the RDA is 22 IU. Nonetheless, subjects in the studies just listed received as little as

400 IU and as much as 25,000 IU of vitamin E. Even at the lower dose, subjects were ingesting more than eighteen times the RDA. This would mean eating about two thousand almonds (a rich source of vitamin E) to obtain the equivalent of what was in one tablet at the lower dose. Our stomachs are only so big for a reason. We aren't meant to eat eight cantaloupes or fourteen oranges or two thousand almonds at one sitting. Taking supplemental antioxidants at doses typically found at GNC or other health food stores is a profoundly unnatural thing to do. You have to give credit to the dietary supplement industry for successfully advertising its products as *natural* when they're anything but.

Ironically, large quantities of supplemental antioxidants might also have a *pro*-oxidant effect. In one study of male smokers, levels of beta-carotene in blood were twelve times higher than in the placebo group. The researchers postulated that at those levels, vitamin A might either inhibit the body's natural antioxidants (e.g., glutathione) or increase the amounts of free radicals that damage cell membranes and DNA, predisposing the user to a greater risk of cancer and heart disease. Large quantities of vitamin E might also displace other fat-soluble vitamins (such as vitamins A, D, and K), disrupting the natural balance and causing an increased vulnerability to oxidative stress.

BY 2010, studies of hundreds of thousands of men and women performed by some of the most prestigious cancer centers in the world had shown that large quantities of supplemental antioxidants increased the risk of heart disease and cancer. Animal studies had shown that supplemental antioxidants accelerated the progression of tumors. And studies of multivitamins, which often contain quantities of antioxidants well above the RDA, had also sounded an alarm. So why don't we know about this? Where is the outcry? When certain pain relievers (e.g., Vioxx) were found to be a rare cause of heart attacks, we knew about it. Indeed, the FDA placed a "black box" label on the package insert, warning patients and doctors about a serious safety problem. Why haven't FDA officials warned consumers that supplemental antioxidants increase the risk of cancer and heart disease? The simple answer is they can't. While the FDA regulates pharmaceutical products, it doesn't, for all practical purposes, regulate dietary supplements. As a consequence, the industry continues to mislead the public.

In 2009, when researchers at the Cleveland Clinic, in collaboration with other cancer centers, found that supplemental vitamin E increased the risk of prostate cancer, a headline appeared in the *Wall Street Journal* asking, "Is This the End of Popping Vitamins?" The article noted that studies showing an increased risk of cancer and heart disease hadn't hurt sales, which had continued to rise. "The thing to do with [these studies] is just ride them out," said Joseph Fortunato, chief executive of GNC Corporation. "We see no impact on our business." Advertising dollars trump scientific studies.

Another recent trend is particularly worrisome. For those who believe ingestion of unnaturally large quantities of vitamins and minerals isn't enough, you can now get them intravenously at IV Bars and a variety of other "intravenous lounges." For example, a Manhattan medical office called Evolved Science offers "Jet Lag Eraser," for use after long flights; a "Detoxification Saline Smoothie," for "improving alertness and mental acuity"; and "Late-Night Rescue," to treat hangovers. "We put together the ideal combination for them to obtain the results they want," said Erika Schwartz, who runs the clinic, "clearer skin, clearer mind, better hair, better nails." Celebrities ranging from Simon Cowell to Rihanna to some of the Real Housewives swear by them.

DESPITE all the hype, not a single national or international organization responsible for the public's health recommends supplemental antioxidants. Nonetheless, more than half the U.S. population takes them.

6

Testosterone for "Low T" Is Dangerous and Unnecessary

The ads, which are directed at men over forty, are hard to resist:

Reduced Sex Drive?

Decreased Energy?

Unwanted Body Changes?

Mood Changes?

Sexual Dysfunction?

Stop Living in the Shadow. If you feel you are experiencing any of the symptoms of Low T, talk to your doctor to see if you should be tested.

Intrigued? Take the "Low-T Test":

Do you have a decrease in libido?

Do you have a lack of energy?

Are you sad or feeling grumpy?

Are you falling asleep after dinner?

Have you noticed a recent deterioration in your ability to play sports?

Are you feeling like a shadow of your former self?

If so, you might have a treatable condition called Low T.

"Low T" stands for low testosterone, the male sex hormone. Testosterone has a variety of functions. It promotes the growth of the testicles and prostate, regulates sex drive (libido), strengthens muscles and bones, and determines fat distribution, the development of body hair, and the production of sperm.

Like many hormones in the body, testosterone levels in the bloodstream are highly regulated. When levels fall, the hypothalamus, which is located at the base of the brain, releases a hormone that stimulates the pituitary gland to release two other hormones that stimulate the testicles to produce more testosterone.

Any disease that affects this hypothalamus-pituitary-testicle axis could result in a critical decrease in testosterone, causing what doctors call "hypogonadism." Causes of hypogonadism include genetic disorders; undescended testicles; testicular injury; infections such as tuberculosis, AIDS, and mumps; tumors of the pituitary gland and hypothalamus; radiation and chemotherapy; drugs such as opiates and antidepressants; and disorders such as alcoholism, obesity, and type 2 diabetes.

Symptoms of hypogonadism aren't subtle. Infertility, muscle wasting, loss of sex drive, loss of body hair, obesity in the abdominal area, erectile dysfunction, decreased penile sensation, reduced ejaculate, reduced energy and stamina, depressed mood, fatigue, hot flashes, sleep disturbance, increased irritability, difficulty concentrating, poor cognition, anemia, difficulty attaining orgasm, and reduced bone density are all caused by a critical lack of testosterone.

Typically, doctors treat male hypogonadism with testosterone tablets, implantable pellets, intramuscular or subcutaneous injections, or transdermal patches, solutions, or gels. Transdermal gels are the most popular.

LOW T, however, is not hypogonadism. It's something else. Beginning at around age thirty, levels of testosterone in men begin to decline, and they continue to decline about 1 percent every year. As a consequence, older men have lower levels of testosterone than younger men. The result is a decrease in sex drive and sexual performance and changes in muscle mass, muscle strength, and bone density. This is a normal physiological process. It's called aging. And it's not a disease. Nonetheless, testosterone makers have made it one, creating an enormous market for their product. Said another way, it's the rare older man who answers "no" to all the questions on the "Low-T Test."

The Low-T craze started in 2002. Prior to that, the gradual lowering of testosterone in older men was called andropause, which was accurate. The prefix *andro-* is derived from the Greek word meaning "man"; the suffix *-pause* from the Latin word meaning "cessation." Andropause is the male equivalent of menopause, which also involves a decrease in a sex hormone (estrogen). By choosing the term *Low T,* however, testosterone manufacturers had created a label that sounded no longer like a natural process, but like a treatable disorder. Sales exploded. In 2002, testosterone makers were selling $324 million worth of their product a year. Ten years later, with an annual advertising budget of $107 million, sales topped $2 billion. Prescriptions for testosterone increased more than threefold. In the United Kingdom, they increased fourfold. When the term *Low T* was translated into other languages, testosterone sales skyrocketed in thirty-seven other countries.

By taking advantage of a loophole in how companies are allowed to promote their products, testosterone manufacturers had succeeded in making aging a disease. By avoiding the names of specific brands (and calling their sales pitches "disease awareness campaigns"), companies could now sell testosterone for the treatment of vague symptoms such as listlessness or increased body fat or declining sexual performance. Also, by encouraging men to take the "Low-T Test," drug companies ensured themselves a ready market. When these nonbranded sales campaigns first launched, some medical specialists complained to the Food and Drug Administration. But they'd complained to the wrong regulatory agency. Only the Federal Trade Commission (FTC) has the authority to regulate this kind of advertising—and the FTC stood back.

In the early 2000s, the Low-T media blitz took the form of television ads, magazine articles written by physicians with ties to the drug industry, and websites explaining how dips in energy, changes in mood or sex drive, and worsening sports performance could all be explained by low testosterone levels. As a consequence, testosterone was sold to healthy men, many of whom never had their testosterone levels checked before receiving a prescription.

In 2013, Larry Dobrow, writing for the trade magazine *Medical Marketing and Media*, sang the praises of this new advertising effort, specifically singling out a company that made testosterone gel. Under the headline "All-Star Large Pharma Marketing Team of the Year," Dobrow wrote, "It might be exaggerating to say that AbbVie—Abbott's new breakout pharma division—overcame a host of challenges to establish AndroGel as the preeminent product in the low-testosterone category. Or maybe not. AbbVie took a taboo topic and, via a cagey media and marketing presence, rendered it less wince-inducing among its target audience. It did so at a time when a number of critics voiced their concerns that the marketing and use of testosterone-boosting products had gotten ahead of the science." Nonetheless, the magazine had named AbbVie executives its "Marketing Team of the Year." Apparently, getting ahead of the science was a good thing.

Lisa Schwartz and Steve Woloshin, codirectors of Dartmouth College's Center for Medicine and the Media, saw the dangers of this new approach to marketing drugs. "We should recognize it for what it is," they wrote, "a mass uncontrolled experiment that invites men to expose themselves to the harms of a treatment unlikely to fix problems that may be wholly unrelated to testosterone levels." The "harms," as it turned out, were far greater than Schwartz and Woloshin could have imagined.

NO one disagreed that lower testosterone, even when associated with aging and not true hypogonadism, was linked to a variety of symptoms. Wasn't it at least possible, manufacturers argued, that replacing testosterone in aging men could have positive effects? This question was easily studied. The results, however, weren't what the companies had anticipated.

Testosterone Doesn't Improve Memory

In 1997, researchers from St. Louis University Health Sciences Center divided thirty-two older men into two groups. One group was injected with 200 milligrams of testosterone every two weeks for a year; the other, a placebo. At the end of the trial, the men were given a series of tests of memory. No differences were detected.

Testosterone Doesn't Improve Cognition

In 2000, researchers at the University of Trier, in Germany, divided thirty older men into two groups. One group was injected with 250 milligrams of testosterone; the other, a placebo. Five days later, cognitive function was determined by a series of spatial and verbal tests. Again, no differences.

In 2006, researchers from the David Geffen School of Medicine, at UCLA, divided thirty-eight men with Alzheimer's disease into two groups. One group was given 75 milligrams of testosterone gel weekly for twenty-four weeks; the other, a placebo. Again, the researchers found no differences in cognitive scores.

Testosterone Doesn't Improve Mood, Quality of Life, or Psychological Well-being

In 2000, researchers at Hartford Hospital divided twenty-two older men into two groups. One group was injected with 200 milligrams of testosterone every two weeks for eight weeks; the other, a placebo. Six weeks after the last dose, both groups were given a series of tests to determine health-related quality of life and psychological well-being. Again, no differences.

In 2005, researchers from the University of Adelaide, in Australia, divided seventy-six older men into two groups. One group received one 80-milligram testosterone tablet twice daily for one year; the other, a placebo. Both groups were then given a series of physical and psychological tests. Again, no differences.

Testosterone Doesn't Improve Depression

In 2001, researchers from Columbia University, in New York City, divided thirty-two men with major depressive disorders into two groups. One group was injected with 200 milligrams of testosterone every week for six weeks; the other, a placebo. The authors concluded that the purported "antidepressant effects of testosterone could not be differentiated from those of placebo."

In 2010, researchers from McLean Hospital in Belmont, Massachusetts, divided one hundred healthy adult men with major depressive disorders into two groups. One group received testosterone gel with the dose constantly adjusted to the level of testosterone in the blood; the other, a placebo. Participants were tested weekly for six weeks for changes in mood. The authors concluded that "Testosterone is not generally effective for depressed men."

Testosterone Doesn't Improve Fatigue, Motor Function, or Bone Mineral Density

In 2006, researchers from the University of Florida in Gainesville divided thirty men with Parkinson's disease into two groups. One group was injected with 200 milligrams of testosterone every two weeks for eight weeks; the other, a placebo. Both groups were given a series of tests to determine mood, cognition, fatigue, and motor function. Again, no differences.

In 2008, researchers from the University Medical Center in Utrecht, the Netherlands, divided 237 healthy older men into two groups. One group was injected with 80 milligrams of testosterone twice daily for six months; the other, a placebo. The authors found that lean body mass increased and fat mass decreased in those who had received the testosterone. However, researchers didn't detect any differences between the two groups in mobility, muscle strength, cognitive function, bone mineral density, or quality of life.

ALTHOUGH testosterone hadn't been shown to work for a variety of symptoms of low testosterone, advocates were certain that at the very least it would help men with erectile dysfunction.

It didn't work out that way.

Testosterone Doesn't Treat Erectile Dysfunction

In 1979, researchers at the Center for Maternal and Child Health in Durham, North Carolina, divided twenty-nine men into two groups. All men had a reduced or nonexistent capacity to have an erection during intercourse (erectile dysfunction). One group received daily injections of 120 milligrams of testosterone for eight weeks; the other, a placebo. The authors concluded that testosterone was "no more effective than placebo in restoring sexual potency to sexually impotent men."

In 1997, researchers from Mount Sinai School of Medicine, in New York City, divided twelve men with erectile dysfunction into two groups. One group received biweekly injections of 200 milligrams of testosterone for six weeks; the other, a placebo. The authors found no differences in sexual desire, masturbation, sexual experiences, sleep erections, penile rigidity, or sexual satisfaction.

In 2012, researchers at Boston University School of Medicine divided 140 men with erectile dysfunction into two groups. One group received sildenafil (Viagra) alone. The other received sildenafil plus 10 milligrams of testosterone gel daily for fourteen weeks. The authors found that while sildenafil caused "a substantial increase in erectile function scores," testosterone offered no additional benefit.

In 2014, in what was probably the most aggressive study of the effect of testosterone on erectile dysfunction, researchers from Heidelberg, Australia, divided eighty-eight obese men with type 2 diabetes into two groups. Both obesity and type 2 diabetes are associated with decreased production of testosterone. One group was injected with 1,000 milligrams of testosterone every six weeks for forty weeks; the other, a placebo. Again, no difference.

THE finding that testosterone didn't treat erectile dysfunction shouldn't have been surprising. Erectile dysfunction is almost always associated with low blood flow to the penis, which is caused by high blood pressure, high cholesterol, and other conditions that narrow blood vessels and reduce flow. That's why sildenafil, which increases blood flow to the penis, worked.

During the past forty years, 156 studies of testosterone have been performed across the globe. In 2018, these trials were reviewed and summarized by researchers from Tulane University School of Medicine, Georgetown University School of Medicine, and the University of South Florida. The researchers concluded that "The prescription of testosterone supplementation for low-T for cardiovascular health, sexual function, physical function, mood, or cognitive function is without support from randomized clinical trials." Further, they found that the continued use or study of testosterone in men who did not have true hypogonadism was unethical: "We identified no population of normal men for whom the benefits of testosterone use outweigh its risks. Given the *known risks* of testosterone therapy and the lack of evidence for clinical benefits in normal men, we do not think further trials of testosterone are necessary."

The "known risks" cited by these reviewers had, unfortunately, only recently become apparent.

FOR decades, clinicians had known that testosterone therapy could cause acne, enlarged prostate, decreased sperm count, and sleep apnea. Indeed, some of these side effects appeared as a warning label on the product. Also, from five studies that had been performed between 1995 and 2010, testosterone was known to increase anger, aggression, and hostility. Because testosterone is a type of steroid, this particularly noxious side effect was called "roid rage."

The poster boy for "roid rage" was Lyle Alzado, a defensive end for the Los Angeles Raiders in the 1980s who injected himself with testosterone for most of his fifteen-year NFL career. His steroid use began in college, when he went from 195 pounds as a freshman to 245 pounds as a sophomore to 280 pounds as a junior to 300 pounds as a senior. Alzado's nicknames included "Rainbow" because of his mood swings and "Three Mile Lyle" because of his volcanic temper. Before Super Bowl XVIII, Alzado vowed to decapitate Washington Redskins quarterback Joe Theismann. During one game, he tore the helmet off an opposing player and threw it at him. The next season, the NFL created the "Alzado Rule," punishing the removal of another player's helmet with ejection and a possible fine. Alzado also once threw a chair at a newspaper reporter and said that his only regret was that he had missed. In 1991, Alzado was arrested for battery of a female police officer who, fearing for her safety, had maced him. "If me and King Kong went into an alley," boasted

Alzado, "only one of us would come out. And it wouldn't be the monkey."

There was one side effect, however, that researchers hadn't anticipated. And its discovery should end the casual use of this drug.

Testosterone Increases the Risk of Stroke and Heart Attack

In 2010, researchers from Boston University School of Medicine divided two hundred men over age sixty-five into two groups. One group received 100 milligrams of testosterone gel daily for six months; the other, a placebo. Researchers wanted to determine whether testosterone increased muscle function. It did. They found that older men who received testosterone had significantly greater improvement in tests of strength. The study, however, had to be stopped prematurely when researchers noticed an abnormally high number of serious side effects in the testosterone group—specifically, increased heart rate, elevated blood pressure, chest pain, heart attacks, heart failure, heart arrhythmia, and stroke. Whereas twenty-three men in the testosterone group experienced one or more of these symptoms, only five in the placebo group experienced them—a significant difference.

Stunned by this result, other groups looked back at their records to see if they could find what the Boston researchers had found.

In 2013, three years after the Boston researchers were forced to suspend their study, investigators from the University of Texas Southwestern Medical Center, in Dallas, examined the medical records of 1,200 veterans who had started testosterone therapy within two years of a coronary angiography, a procedure to determine whether the arteries that supply the heart are partially blocked. Researchers then compared these men with 7,500 men who also had undergone a coronary angiography but hadn't later received testosterone. Similar to the Boston study, the researchers found that testosterone treatment increased the risk of strokes and heart attacks.

That same year, researchers from the University of Hong Kong reviewed twenty-seven studies involving three thousand older men who had received testosterone and compared them with men of the same age who hadn't received the drug. They found that those who had received testosterone were twice as likely to have heart artery disease, arrhythmias, heart failure, and stroke. Perhaps most discouraging, researchers found that the risk of heart problems varied with the source of funding: "In trials not funded by the pharmaceutical industry[,] the risk of a cardiovascular-related event on testosterone therapy was greater than in pharmaceutical industry funded trials."

One year later, in 2014, researchers from the University of California, Los Angeles, and the National Institutes of Health, in Bethesda, examined the medical records of 56,000 men who either had or hadn't received testosterone. They found that in men over age sixty-five, the risk of a heart attack within ninety days of receipt of testosterone was greater than twice that of men who hadn't received the drug; for those over seventy-five, the risk was almost four times greater.

THEN the FDA stepped in. On January 31, 2014, in a warning letter on its website, it tried to put an end to the myth of Low T, arguing that only men with true hypogonadism should be treated with testosterone: "Testosterone is FDA-approved as replacement therapy only for men who have low testosterone levels due to disorders of the testicles, pituitary gland or brain that cause a condition called hypogonadism. Examples of these disorders include failure of the testicles to produce testosterone because of genetic problems, or damage from chemotherapy or infection. *However, FDA has become aware that testosterone is being used extensively in attempts to relieve symptoms in men who have low testosterone for no apparent reason other than aging*" (emphasis mine).

The FDA then did something that it hoped would put the brakes on this ill-conceived industry, demanding a change in the labeling of testosterone: "Based on our findings, *we are requiring labeling changes for all prescription testosterone products to reflect possible increased risk of heart attacks and strokes associated with testosterone use*" (emphasis mine).

On February 7, 2014, the Endocrine Society, a professional organization in the United States that represents doctors most likely to prescribe testosterone, sent a letter to its members. After the fourth study in five years found that testosterone was far more dangerous than doctors had realized, they knew that a line had been crossed. At the very least, doctors had to let their patients know what was happening. Noting that the number of men taking testosterone in the United States since 2000 had almost quadrupled, they wrote, "The Endocrine Society believes that patients should be made aware of the potential risk of [heart artery] disease in middle-aged and older men who are taking or considering taking testosterone therapy."

Unfortunately, the Endocrine Society's warnings and the FDA's labeling changes have done little to discourage men from taking testosterone for a disorder that isn't a disorder. Millions of men still spend hundreds of dollars a month for a product that hasn't fulfilled its promise and, worse, puts them at risk of a potentially fatal side effect.

PERHAPS the most discouraging part of this story is that we have failed to learn from history and, as a result, have been condemned to repeat it.

The predominant sex hormone in women is estrogen, which is produced in the ovaries. Whereas levels of testosterone gradually decrease as men age, for women, the decline is more sudden, typically occurring around fifty years of age. The sudden decline in estrogen is called menopause, which is defined as the absence of a menstrual period for twelve consecutive months. (The prefix *meno-* is derived from the Greek word meaning "month.")

Like andropause, menopause is a natural biological process. And like andropause, menopause is accompanied by a variety of symptoms, including vaginal dryness, hot flashes, chills, night sweats, sleep problems, mood changes, weight gain, slowed metabolism, thinning hair, dry skin, and loss of breast fullness.

In 1942, the FDA approved estrogen for the treatment of hot flashes associated with menopause. Beginning in the 1950s, however, health advocates started to raise concerns that estrogen might be doing more harm than good. These warnings were largely ignored. Then, in 1966, estrogen therapy took off. The launching pad was a book written by Robert A. Wilson titled *Feminine Forever*. Similar to the creation of the term *Low T*, Wilson argued that menopause wasn't a natural process, but a preventable disease—and estrogen was the cure. Wilson wrote that with estrogen, women's "breasts and genital organs will not shrivel. She will be much more pleasant to live with and will not become dull and unattractive."

Feminine Forever became a national bestseller. Women flocked to their gynecologists demanding estrogen. Although it had been approved only for the treatment of hot flashes, doctors began prescribing it for all manner of symptoms associated with menopause. By 1992, estrogen was the most prescribed drug in the United States. By 1997, sales exceeded $1 billion.

Then it all came crashing down. Between 1993 and 1998, researchers from forty U.S. medical centers recruited 17,000 postmenopausal women to participate in a randomized, controlled trial of estrogen called the Women's Health Initiative. The researchers found that, similar to men taking testosterone, women taking estrogen had an increased risk of strokes, blood clots, and heart attacks. They also had an increased risk of uterine cancer.

ANDROPAUSE and menopause aren't diseases, but they aren't alone in being declared diseases by doctors and pharmaceutical companies. For example, moods such as sadness aren't moods anymore; they're affective disorders requiring medicines. Other examples abound. Chronic fatigue syndrome and chronic pain syndromes are often a somatization of life stresses or family dysfunction. Nonetheless, most of those who suffer from them are treated with polypharmacy, not counseling.

Aging is a natural process. Taking large quantities of sex hormones isn't. Although some would argue that the phrase "aging gracefully" is an oxymoron, attempts to turn back the clock have come with a price, and it's not insignificant.

7

Baby Aspirin Doesn't Prevent First Strokes or First Heart Attacks

Aspirin is the most commonly used drug in the world. Nonetheless, for the past five thousand years, no product has been more celebrated, more vilified, more embraced, or more feared. How did this remarkable drug achieve such a central place in the pantheon of medicine?

Aspirin is derived from the bark and leaves of the willow tree. It can also be found in the meadowsweet flower. Beginning around 3000 BC, ancient civilizations such as the Egyptians and Sumerians chewed willow bark to relieve aches and pains. In 400 BC, Hippocrates recommended willow leaf tea for women during childbirth. But it wasn't until 1828 that Joseph Buchner, a professor of pharmacy at Munich University in Germany, successfully extracted the active ingredient from willow, producing bitter-tasting yellow crystals that he called salicin (from *Salix*, the genus name for the willow tree).

The critical discovery came in 1897. That's when Felix Hoffmann, working for a small pharmaceutical company in Bavaria, purified salicin from meadowsweet flowers and chemically modified it (by acetylation). To name the drug, he chose an *a* to denote the acetyl group and combined it with *spir*, from the Latin name of the meadowsweet flower (*Spiraea ulmaria*), and *in*, from the word *salicin*. The company he worked for was Bayer. The product was Bayer Aspirin.

Ten days after Hoffmann added one acetyl group to salicin, he added two acetyl groups to morphine, a well-known pain reliever derived from the opium plant. The chemical name of his new drug was diacetylmorphine. Hoffmann hoped that by acetylating morphine, he had made it less addictive. Bayer named the drug for its heroic properties. For twenty years, diacetylmorphine was sold over the counter. However, it was found to be far more addictive than the company had promised. In 1924, the United States banned Bayer Heroin.

BAYER Aspirin came onto the market with a bang. Within a few years, more than 160 scientific papers had been published extolling its virtues as a painkiller and antifever medicine (antipyretic). In 1950, aspirin entered the *Guinness Book of World Records* as the most frequently used drug in the world.

But problems had already started to arise.

In 1932, with the invention of the gastroscope, aspirin's corrosive effects on the lining of the stomach—it caused ulcers and inflammation (gastritis)—were becoming readily apparent. This left it vulnerable to the discovery of painkillers that weren't as harsh, such as acetaminophen (Tylenol), which was invented in 1948, and ibuprofen (Motrin, Advil), which was invented in 1961. Worse, two years after ibuprofen was invented, aspirin was found to be the cause of a rare but occasionally fatal disorder called Reye syndrome, symptoms of which included swelling of the liver and brain. Reye syndrome occurred most commonly in children and teenagers. By the mid-1980s, aspirin was essentially banned for use in children.

With the availability of other painkillers, the side effects of ulcers and gastritis, and the relatively rare but occasionally fatal problem of Reye syndrome, it looked like aspirin had finally outlived its usefulness. Then the drug reinvented itself.

In 1971, John Vane, a British pharmacologist, discovered that aspirin inhibited an enzyme that promoted inflammation. For this discovery, Vane was awarded the Nobel Prize in 1982. Two years later, he was knighted.

A few years before Vane won the Nobel Prize, two American physicians found that aspirin could also inhibit platelets, small cells in the bloodstream necessary for clotting. As it turned out, the same enzyme that Vane had shown promoted inflammation also promoted platelet aggregation. Aspirin was back, reborn as an antiplatelet drug that could now prevent strokes and heart attacks.

Here's how it worked.

Arteries are lined with endothelial cells. When these cells are damaged, inflammation results. If the damage is severe, platelets stream to the site of inflammation, aggregate, and form a clot, blocking the artery. If a clogged artery supplies the heart, patients suffer a heart attack. If it supplies the brain, they suffer a stroke.

Several processes can disrupt the endothelial cells that line the arteries, including high blood pressure (hypertension), high levels of glucose in the bloodstream (diabetes), high levels of low-density lipoprotein (bad) cholesterol (hyperlipidemia), and high levels of cell-damaging molecules called free radicals, caused by cigarette smoking.

Because diabetes, hypertension, hyperlipidemia, and cigarette smoking are common, strokes and heart attacks are common. Every year in the United States about 370,000 people die from heart attacks, the leading cause of death for both men and women. And every year, about 800,000 people suffer stroke.

Because platelets are directly responsible for clogging arteries, it stood to reason that aspirin, which inhibits the enzyme that allows platelets to aggregate, could prevent strokes and heart attacks. The first studies to prove that aspirin worked in this way were performed in people who had already had a stroke or heart attack, putting them at highest risk of suffering another one.

Aspirin Prevents a Second Stroke or Second Heart Attack

In 1974, researchers from Cardiff University in Wales divided 1,200 people under age sixty-five who had previously suffered a heart attack into two groups. One group received 300 milligrams of aspirin every day; the other, a placebo. After six months, aspirin had caused a 12 percent reduction in mortality. After twelve months, a 25 percent reduction. This was the first evidence that aspirin not only lessened pain and fever but also saved lives.

In 1987, researchers from Radcliffe University in Oxford, England, divided 17,000 patients who had recently suffered a heart attack into two groups. One group received 162 milligrams of aspirin every day; the other, a placebo. Thirty-five people died in the placebo group and 16 in the aspirin group. Again, aspirin had saved lives.

In 2002, these same researchers reviewed 287 studies involving 135,000 patients who had previously suffered a stroke or heart attack and had subsequently received 75 to 150 milligrams of aspirin every day for two years. This lower dose of aspirin is commonly referred to as "baby" aspirin. The researchers found that aspirin reduced the risk of heart attacks by 33 percent, of strokes by 25 percent, and of overall mortality from either strokes or heart attacks by 17 percent.

In 2009, these same researchers reviewed sixteen more studies and found that aspirin reduced the risk of a second stroke or second heart attack by about 20 percent.

Because the results of these studies were clear, everyone who has suffered a stroke or heart attack is now recommended to receive daily low-dose (baby) aspirin. But what about people who haven't suffered a stroke or heart attack but are still at high risk, like people with diabetes or hypertension or hyperlipidemia or who are elderly or who smoke? Would they also benefit from daily aspirin?

Initially, it appeared that they would.

IN 1989, researchers from Harvard Medical School divided 22,000 people at high risk of strokes or heart attacks, but who had never had one before, into two groups. One group received 325 milligrams of aspirin every day for five years; the other, a placebo. The researchers found that those taking aspirin had a 44 percent reduction in their risk of a heart attack. Benefits were greatest in participants who were more than fifty years old.

The Harvard study launched the use of daily aspirin to prevent heart attacks in millions of people at high risk. Subsequent studies, however, failed to support these dramatic findings.

Aspirin Doesn't Prevent a First Stroke or First Heart Attack, Even in Those at Highest Risk

In 2008, researchers from Kumamoto University in Japan divided 2,500 patients with diabetes into two groups. One group received 81 milligrams of aspirin every day for four years; the other, a placebo. The authors found no differences in the incidence of fatal and nonfatal heart attacks or strokes.

In 2010, researchers from the University of Glasgow, in Scotland, divided 29,000 older men and women into two groups. One group received 100 milligrams of aspirin every day; the other, a placebo. Eight years later, the researchers found no differences in the incidence of strokes or heart attacks.

In 2014, researchers from Waseda University in Tokyo divided 14,000 patients with diabetes, hypertension, or hyperlipidemia into two groups. One group received 100 milligrams of aspirin every day for five years; the other, a placebo. The researchers found no differences in the incidence of heart attacks, strokes, or overall mortality. Once again, baby aspirin hadn't saved the lives of those at highest risk.

Despite these studies, millions of high-risk patients in the United States are still taking low-dose aspirin. Additional studies reported between 2016 and 2019 should end this practice. Because, as we have now learned, the risks of taking aspirin aren't small. And although the term *baby aspirin* is somewhat comforting, it shouldn't be.

In 2016, investigators from Oxford, England, completed a seven-year study in which they randomly assigned 15,000 adults with diabetes into two groups. One group received 100 milligrams of aspirin every day; the other, a placebo. The rate of strokes or heart attacks was 9.6 percent in the placebo group and 8.5 percent in the aspirin group—a 12 percent reduction. This benefit, however, came with a high price. The incidence of bleeding in the aspirin group was 29 percent greater than in the placebo group. Some of these bleeding events were minor, such as nosebleeds or blood in the urine. Others were more severe, such as intestinal bleeding requiring hospitalization, sight-threatening bleeding in the eye, and intracranial bleeding between the brain and skull. Also, the reduction in heart attacks found in this study was a little misleading. Modern-day screening techniques to determine heart attacks include the sensitive measurement of heart-specific enzymes in the bloodstream, which can be released when the heart is only slightly damaged. If you exclude these transient, small episodes, the risks of aspirin in this trial clearly outweighed its benefits. The authors concluded that "The absolute benefits [of aspirin] were largely counterbalanced by the bleeding risk."

In 2018, researchers from Monash University, in Melbourne, Australia, reported a five-year study in which they divided 19,000 people seventy years of age or older into two groups. One group received 100 milligrams of aspirin every day; the other, a placebo. The authors found no differences between the two groups in the incidence of heart attacks, strokes, dementia, persistent physical disability, or death. And like the United Kingdom researchers, they found an unacceptably high incidence of serious bleeding in the aspirin group.

In 2019, researchers from the Faculty of Medicine at Imperial

College London reviewed thirteen more studies of aspirin for the prevention of first strokes or heart attacks in 164,000 people at high risk. They, too, found that any possible benefits of aspirin were outweighed by its risks.

IF aspirin doesn't convincingly prevent first strokes or first heart attacks in people at risk, why did the 1989 study at Harvard Medical School—the one that found a 44 percent reduction in the risk of heart attacks and launched the recommendation to take a daily aspirin—show that it did? The answer is that much has changed during the past thirty years. This Harvard study was done at a time when cigarette smoking was common, blood pressure control was less than optimal, and aggressive treatment of bad cholesterol was rare.

What we can now conclude more than one hundred years after it was first synthesized is that aspirin benefits people who have already had a stroke or heart attack. For these people, the risk of a second stroke or second heart attack is so high that aspirin's benefits outweigh its significant bleeding risks. Such is not the case for people at risk for a stroke or heart attack but who have never had one. In these cases, the risk of bleeding outweighs any possible benefits. In the end, the best bet for those at highest risk is diet, exercise, and smoking cessation; and for those with hyperlipidemia, statins.

THEN, to the surprise of everyone, aspirin reinvented itself one more time.

Aspirin Prevents Colon Cancer

In 2010, researchers from the University of Oxford, in collaboration with researchers from several other European countries, reviewed the records of 14,000 patients who either had or hadn't received aspirin to prevent strokes and heart attacks. They found, much to their surprise, that the aspirin group had a 24 percent reduction in the incidence of colon cancer! Surely this must be a fluke, some argued, a spurious association. Why would aspirin prevent colon cancer? This study prompted other researchers to look back at their records to see if they could find what the Oxford researchers had found.

In 2015, investigators from the Danish Cancer Society, in Copenhagen, Denmark, reviewed the records of 113,000 people who had or hadn't taken aspirin to prevent heart attacks. Again, those who had taken aspirin had a 27 percent reduction in the incidence of colon cancer.

Also in 2015, investigators from the Fred Hutchinson Cancer Research Center, in Seattle, Washington, reviewed ten more studies involving 17,000 patients from the United States, Canada, Australia, and Germany who either had or hadn't taken aspirin between 1976 and 2011. They, too, found that aspirin had reduced the risk of colon cancer—this time by 29 percent.

All these studies were remarkably consistent. How could this happen?

To understand how aspirin prevents colon cancer, we need to go back to the work of John Vane, the British pharmacologist who discovered how aspirin worked. The enzyme responsible for inflammation and platelet aggregation (the one that Vane found was inhibited by aspirin) is called cyclooxygenase-2 (or COX-2). This same enzyme is also found in colon cancer cells but not normal colon cells. Researchers at Massachusetts General Hospital, part of Harvard Medical School, later showed that aspirin inhibits the growth of colon cancers that make large quantities of COX-2 but not those that don't make the enzyme. Doctors, who for decades had been giving aspirin to prevent heart attacks or strokes, had also unknowingly been preventing certain colon cancers.

Given that every year in the United States fifty thousand people die from colon cancer, wouldn't it make sense to take a daily aspirin to prevent it? The answer to this question is "no." For a few reasons.

Aspirin reduces the risk of colon cancer by about 25 percent. Daily aspirin therapy could, then, theoretically save about twelve thousand lives a year. However, not all colon cancers make COX-2. About a third of these cancers don't express much COX-2 at all, rendering them resistant to aspirin. For this reason, the figure of twelve thousand, representing the lives that we've just theoretically saved, is now down to about eight thousand.

Most important, other strategies are available to prevent colon cancer. Specifically, the American Cancer Society recommends colonoscopies every ten years beginning at age fifty. More recently, the Society recommended Cologuard, which tests for the presence of cancer-associated DNA in stool, to be performed every three years. If everyone followed these recommendations, deaths from colon cancer would decrease dramatically. People with colon cancers that are detected early by colonoscopies or Cologuard (so-called stage 1 cancers) have a 95 percent chance of survival.

But what about colon cancers that despite careful screening aren't detected until they are at stage 4, when the chance of survival drops to 10 percent? Wouldn't it make sense for people to take daily aspirin to prevent these hard-to-detect types of colon cancer? If aspirin were risk-free, that answer might be "yes." But it's not. In people under seventy-five, the incidence of fatal bleeding events caused by aspirin is 0.33 percent. Although 0.33 percent doesn't sound like a lot, when you're talking about millions of people, it is. For example, for every million people under age seventy-five taking aspirin daily, about 3,300 will die from intestinal or intracranial bleeding caused by aspirin. For people over age seventy-five, the incidence of fatal bleeding is 2.2 percent. For those over age eighty-five, it shoots up to an astronomical 4.1 percent.

FOR people who have suffered a stroke or heart attack, aspirin is clearly of value in preventing a second one. But that's it. The risk of severe, disabling, or fatal bleeding caused by aspirin now outweighs its benefits in other situations. Although low-dose aspirin is often benignly referred to as "baby" aspirin, babies are rarely given this drug. It's just too dangerous. The same holds true for most adults.

PART III

Food and Over-the-Counter Products

8

Embrace Allergenic Foods for Infants

In 2004, a nurse in a San Francisco elementary school insisted that every child thoroughly scrub their hands and submit their packed lunches for inspection because a five-year-old with a peanut allergy attended the school.

In 2006, a Connecticut town official cut down three sixty-foot-tall hickory trees when a woman complained that they posed a threat to her grandson, who had a nut allergy.

In 2008, a Massachusetts school district ordered the immediate evacuation of a school bus packed with ten-year-olds after a stray peanut was found on the floor.

Peanut phobia reached its illogical peak when a Canadian teenager supposedly suffered a fatal allergic reaction after being kissed by her boyfriend, who, nine hours earlier, had eaten a peanut butter sandwich. In truth, the girl's fatal asthma attack was unrelated to the kiss. Although the "peanut butter kiss of death" was a media invention, our fear of food allergies is well founded.

Every year, about 15 million Americans suffer from food allergies; 3 million are children. And every year, about 9,500 of these children are admitted to the hospital with hives, itching, swollen lips and tongue, difficulty breathing, wheezing, fainting, vomiting, abdominal pain, low blood pressure, decreased oxygen to the brain and other vital organs, and occasionally shock and death—all caused by an allergic reaction to something they had just eaten.

The single best treatment for a severe allergic reaction is a shot of epinephrine. Unfortunately, because no one likes giving or getting a shot, epinephrine is not always used frequently enough or early enough during these episodes. Medications given by mouth (e.g., antihistamines, steroids, and drugs that relax airway muscles) offer little benefit. Also, although epinephrine can effectively *treat* severe food allergy reactions, no specific drug has ever been found to *prevent* them.

The foods that are most likely to cause allergies are milk, eggs, peanuts, tree nuts, soy, wheat, fish, and shellfish. Among these eight foods, however, one stands above the rest, causing more fatal or near-fatal allergic reactions than any other: peanuts. Every year in the United States, between seventy-five and one hundred people die from peanut allergies.

For those who are allergic, peanuts are hard to avoid. They're placed in salads, sprinkled onto frozen yogurt, crumbled onto a variety of Chinese and Thai dishes, and added as a hidden ingredient to a vast array of other foods. Our fear of peanuts has reached the point where many schools no longer allow parents to send in home-baked foods, unsure of what they contain. Although children often outgrow allergies to milk or eggs by the time they go to school, peanut allergies last well into adulthood. Perhaps worst of all, the incidence of peanut allergies in Western countries has doubled during the past ten years.

Although every year about 100 people are struck and killed by lightning, 1,300 are killed in gun accidents, and 45,000 die in car accidents, food allergies evoke a unique kind of horror. Dr. Robert Wood, a pediatric allergist at Johns Hopkins Hospital, summed it up best. We are now at the point, argued Wood, where "a Snickers bar fifty feet away is a lethal weapon."

Is there any way to solve this growing problem?

Initially, advisory bodies in the United Kingdom in 1998 and the United States in 2000 reasoned that avoidance of peanut-containing foods early in life would lessen the incidence of peanut allergies. They recommended that women avoid eating peanuts during pregnancy and breastfeeding and avoid giving peanut-containing foods to children until they were at least three years old. The problem with these recommendations was that they didn't work. The incidence of food allergies only worsened. In 2008, confronted with overwhelming evidence that their recommendations had been useless, both these advisory groups withdrew them.

Then researchers in the United Kingdom performed a groundbreaking and, given the results, shocking study. They asked parents of five thousand Jewish children in the United Kingdom about their dietary habits and compared the results with five thousand Jewish children in Israel. Although these two groups had similar genetic backgrounds, the incidence of peanut allergies was 1.85 percent for children in the United Kingdom but only 0.17 percent for children in Israel—a tenfold difference. Researchers then asked about the introduction of peanuts into the diet, assuming that Israeli parents were better at avoiding peanuts than their UK counterparts. In fact, the opposite was true. The researchers found that babies in Israel ate about 7 grams of peanut protein a month beginning at seven months of age compared with 0 grams for UK infants. Israeli mothers were feeding their babies a popular peanut butter–flavored snack made in Israel called Bamba. Was it possible that early exposure to peanuts actually *reduced* the risk of peanut allergies? These same researchers decided to put it to the test. In 2015, they published their results in the *New England Journal of Medicine*. They called it the LEAP trial, for "Learning Early About Peanut Allergies."

Between December 2006 and May 2009, UK researchers studied 640 infants between four and eleven months of age who had eczema or egg allergies or both. They reasoned that these children were most likely to develop peanut allergies. Then they randomly assigned them to one of two groups. One group consumed 6 grams of peanut protein per week until they reached age five; the peanut-containing snack used in this study was either Bamba or soft, non-crunchy store-bought peanut butter. The other group was instructed to completely avoid all peanut-containing foods. At the end of the study, 13.7 percent in the peanut-avoidance group and 1.9 percent in the peanut-consumption group were allergic to peanuts—an 86 percent difference. In other words, eating peanuts early in life *decreased* the risk of peanut allergies later in life.

After the LEAP trial was over, the researchers wanted to find out what would happen if these peanut-consuming children stopped eating peanuts for one year. Would they still be resistant to peanut allergies? Or did they need to keep eating the peanuts to avoid the allergies? They found that resistance to peanut allergies persisted; 21.5 percent in the peanut-avoidance group and 2.4 percent in the peanut-consuming group were allergic to peanuts. Even after avoiding peanuts for a year, tolerance to peanuts remained. Four years of eating peanuts was enough to induce a long-lived reduction in peanut allergies.

As a consequence of these studies, in 2017, an expert panel reporting to the National Institutes of Health issued the following guidelines:

1. Parents of children with severe eczema, egg allergies, or both should introduce peanut-containing foods as early as 4 to 6 months of age to reduce the risk of severe peanut allergy.

2. The total amount of peanut protein to be regularly consumed per week should be approximately 6 to 7 grams divided over three or more meals.

3. For infants without eczema or any known food allergies, parents can give potentially allergenic foods early in life.

In 2018, another study proved that children were less likely to develop peanut allergies if their mothers ate peanuts while breast-feeding.

Ironically, the best way to avoid peanut allergies later in life is to embrace peanuts early in life. It will be interesting to see whether early exposure to other highly allergenic foods works just as well.

9

The False Security of Sunblock

Only mad dogs and Englishmen go out in the midday sun.

—NOEL COWARD

First, let's define terms. Sunblock doesn't exist. If you want to block the sun, stay indoors or wear protective clothing. The better term is *sunscreen*, which accurately conveys the concept of screening, not blocking, the sun's harmful rays. People use sunscreens to prevent sunburn and skin cancer. The question is: Do they work?

Skin cancer is the most common cancer in white populations. Three different types have been described: squamous cell (which involves the most superficial layer of the skin), basal cell (which involves deeper layers of the skin), and malignant melanoma (the only form of skin cancer that can spread to other organs, such as the lungs, liver, and brain).

The most common skin cancer is basal cell; about one out of every seven people will develop this cancer in their lifetime. Next is squamous cell, which affects one out of every twenty people. Both these cancers are treated surgically. While often disfiguring, they're rarely fatal. Least common, with a lifetime risk of one in seventy-six, is malignant melanoma. Unlike basal and squamous cell cancers, malignant melanoma can be fatal; 15 percent of people with this disease die from it.

Perhaps most concerning is that despite greater public awareness, the incidence of malignant melanoma is increasing. According to the National Cancer Institute, between 1975 and 2015, the incidence of melanoma had tripled and the death rate had doubled. In 2018, 91,000 Americans were diagnosed with malignant melanoma.

Those at greatest risk for skin cancer share several features. They are more likely to have freckles, fair skin, red or blond hair, blue or green eyes, a large number of moles, and a close relative with the disease. But the single greatest risk factor for skin cancer is exposure to ultraviolet radiation from the sun. The greater the exposure, the greater the risk.

THE sun emits ultraviolet (UV) radiation in three different wavelengths: UVA (320–400 nanometers [nm], which is one-billionth of a meter), UVB (280–320 nm), and UVC (100–280 nm). UVC radiation is completely absorbed by ozone in the stratosphere, so it never reaches the earth's surface. UVB radiation, which peaks around midday, comprises only about 5 percent of the sun's total UV radiation, but it's by far the deadliest. UVB radiation is one thousand times more efficient at producing sunburn than UVA radiation and was at one time considered the only type of UV radiation that caused cancer. For that reason, many early sunscreens protected against UVB but not UVA radiation. Now, however, it has become clear that UVA radiation, which occurs equally throughout the day, can also cause cancer. As a consequence, most sunscreens today protect against both UVB and UVA radiation.

How do we know that UV radiation causes skin cancer?

The first proof came from studies in experimental animals in the 1980s. Mice were put in a box and exposed to UVA or UVB radiation for one, four, or twelve hours. Mice exposed for four or twelve hours developed the same number of malignant tumors at the same rate—the additional eight hours didn't make a difference. Those exposed for only one hour eventually developed the same number of tumors; it just took a little longer. Researchers studying mice found that both UVA and UVB radiation caused cancer. The mechanisms by which they did it, however, were different. Whereas UVB radiation directly damaged DNA, UVA radiation caused oxidative stress, which created free radicals that eventually damaged DNA. This opened the door to the possibility that supplemental antioxidants might lessen the risk of skin cancer.

Unlike mice, scientists can't put people in a box and expose them to different levels of UVA and UVB radiation to determine the effects. As it turns out, they didn't need to. A lot of people do that on their own—in tanning beds. Tanning bed studies mimicked mouse studies. Researchers found that the risk of melanoma increased with the number of tanning bed sessions, the number of hours per session, and the number of years of use. And it didn't matter how young or how old people were when they started or what type of tanning bed they used; everyone was at increased risk of skin cancer. The lessons from these studies were clear: safe tanning beds don't exist, and no one is too young or too old to get skin cancer.

Given that ultraviolet radiation causes skin cancer and that sunscreens lessen UV radiation, researchers assumed that studies in people would clearly show that sunscreens prevented cancer. As it turns out, these studies have been far from reassuring.

SUNSCREENS are numbered based on the degree to which they prevent sunburn. Let's say that it takes a certain dose of UV radiation to produce a sunburn. If a sunscreen is labeled as sun protection factor (SPF) 15, that means that it would take a dose of UV radiation fifteen times greater to produce the same sunburn. Said another way, SPF15 screens out 93 percent of the sun's UVB rays; SPF30 screens out 97 percent; SPF50 screens out 98 percent; and SPF100 screens out 99 percent. None—and this is part of the problem—screens out 100 percent.

Researchers have now studied whether sunscreens of various SPFs prevent skin cancer. In 1999, they divided 1,600 adults in

Queensland, Australia, where sunshine is plentiful, into four groups. One group received SPF15 sunscreen plus beta-carotene (an antioxidant and vitamin A precursor). Every day, this group applied sunscreen to the head, neck, arms, and hands, and every day, they took 30 milligrams of beta-carotene. To determine the relative impact of sunscreen and beta-carotene, another group received sunscreen alone, and a third group received beta-carotene alone. Those in the fourth group were allowed to use sunscreen at their own discretion. Every three months, for about five years, dermatologists examined the participants. Researchers found that daily application of sunscreen alone prevented squamous cell but not basal cell cancers. Beta-carotene offered no additional benefit.

Why did sunscreen prevent one type of skin cancer and not another? The answer lies in where these two different types of skin cells are located. Squamous cells, which reside in the most superficial layer of the skin, don't reproduce themselves very much. Therefore, it's much harder to alter their DNA and much easier to prevent cancer. Basal cells, which are located in deeper areas of the skin, reproduce themselves constantly. Therefore, it's much easier to mutate their DNA and much harder to prevent cancer.

But what about melanoma, the form of cancer that is most likely to be fatal? Studies of the impact of sunscreen on melanomas have been more confusing than clarifying.

In 1999, a researcher from Harvard Medical School reviewed existing evidence. He found two studies that showed that sunscreen prevented melanoma; three that showed that it didn't make a difference; and six that showed that sunscreen actually *increased* the risk. These six studies gave birth to the notion that sunscreen could actually cause melanomas, which precipitated a mild panic among sunbathers and dermatologists.

Researchers argued about the results of these studies. If UV radiation was a risk factor for melanoma, and sunscreen reduced UV radiation exposure, why were the results inconclusive? Several possibilities were discussed. Most obvious, it is virtually impossible to control for sunscreen use in all participants. While researchers could control for how much and what type of sunscreen was used in the "sunscreen" group, they couldn't ask the control group not to use any sunscreen. That would have been unethical. So, control group participants used sunscreen whenever they wanted. Also, unlike squamous cell and basal cell cancers, where UV radiation appears to be the only risk factor, melanoma is more complicated. Squamous cell and basal cell cancers always appear on sun-exposed areas; melanoma, however, can appear on areas that have never been exposed to the sun. Also, melanoma can occur in people who spend almost all their time indoors.

The researcher who finally figured out why sunscreen in people had actually *increased* the risk of melanoma was part of a team of European scientists in Italy, France, and Belgium. His name is Philippe Autier.

In 2000, Autier and colleagues published the results of a series of studies they had begun in 1998. The researchers divided fifty-eight European vacationers between ages eighteen and twenty-four into two groups. One group received SPF10 sunscreen and the other SPF30. Participants were asked to carry around a dosimeter during their sunbathing sessions. This allowed the researchers to determine exactly how much UVA and UVB radiation each of the groups had received. Autier found that those who used SPF30 stayed out longer in the sun than those who used SPF10, presumably because they thought they were less likely to get a sunburn with the higher level of protection. Also, SPF30 users were more likely to sunbathe around midday, when UVB radiation was greatest. SPF10 users, however, because they knew they were more likely to get a sunburn than the SPF30 group, were more likely to sunbathe later in the afternoon, when sunlight was less intense. The authors concluded that "Sunscreens used during sunbathing tended to increase the duration of exposures to doses of ultraviolet radiation below the sunburn threshold"—but, as it turns out, not below the cancer-causing threshold.

Up to this point, dermatologists had assumed that sunburns led to melanoma and that the greater the number of sunburns over the course of one's life the greater the risk of this cancer. But Autier and his colleagues had raised the possibility that you didn't have to be sunburned to develop melanoma. The mechanism by which UV radiation could cause cancer without causing sunburn was determined many years later by a group of researchers working at the Cancer Research Institute in Manchester, England. Richard Marais was the lead investigator.

Marais studied 110 mice that were genetically susceptible to melanoma. Half the mice were protected with SPF50; the other half were unprotected. Mice were then exposed to a dose of UV radiation that was equivalent to spending an hour gardening in the sun in southern England. Marais found that SPF50 delayed but didn't prevent UV radiation–driven melanoma. Unprotected mice developed tumors on average five months after UV exposure; all developed them by seven months. SPF50-protected mice developed tumors on average seven months after exposure; all developed them by fifteen months. So, while SPF50 sunscreen clearly delayed the onset of melanoma, it didn't prevent it. Marais and his colleagues also showed exactly how UVB radiation caused melanoma. Specifically, by disabling a so-called guardian gene (TP53), which mice and people use to protect themselves against radiation-induced melanomas. Up to this point, the TP53 guardian gene was not believed to be important in protecting against melanomas. Marais showed that it was critically important.

Dr. Julie Sharp, another cancer researcher in the United Kingdom, summed it up best: "People tend to think they're invincible once they've put [sunscreen] on and end up spending longer out in the sun, increasing their overall exposure to UV rays." Sunscreens, as it turned out, had provided a false sense of security. People had assumed that if they weren't getting sunburned, they wouldn't get cancer, but they were wrong.

Given what we now know about the sun's harmful rays, skin cancer can best be prevented by heeding the following advice:

1. The first line of defense is to avoid the sun.

2. The second line of defense is to wear protective clothing like broad-brimmed hats, T-shirts, and pants.

3. The third line of defense is to use sunscreen with an SPF of 30 or greater that protects against both UVA and UVB radiation (read the label); apply at least 1 ounce of sunscreen over your entire body thirty minutes before going outside; generously reapply sunscreen every two hours or after swimming or excessive sweating; wear UV glasses; and seek out the shade between 10:00 am and 4:00 pm, when UVB radiation is greatest.

Not surprisingly, few people follow these recommendations. Worse, sunscreens with high protection factors cause people to stay out in the sun longer and during times when UVB radiation is greatest, in the false belief that if they don't get a sunburn then they won't get melanoma.

10

Avoid Reflux Medicines for Fussy Babies

Two facts about babies are inviolate: (1) babies cry, and (2) babies spit up.

Many parents believe that babies cry *because* they spit up. They reason that acid, which is produced in the stomach, is irritating. To lessen the pain, and therefore lessen the crying, parents ask for and occasionally demand medicines to neutralize stomach acid or increase stomach motility: "reflux medicines." As a consequence, reflux medicines are second only to antibiotics as the most commonly prescribed drugs for infants.

Spitting up, or gastroesophageal reflux (GER), is a normal physiologic process that occurs in about two-thirds of infants and is the topic of discussion in about one-quarter of all pediatric visits. It's not hard to understand why. Infants eat all-liquid diets (either formula or breast milk); they usually lie flat on their backs after eating (which makes it easy for stomach contents to travel back up into the esophagus and mouth); and the muscle between the lower end of the esophagus and the stomach, which tightens after food has entered the stomach, isn't fully developed at birth. This muscle, called the lower esophageal sphincter, matures by the end of the first year, which explains why virtually 100 percent of infants will stop spitting up by the time they are one year old. (So hang in there.)

GER becomes gastroesophageal reflux *disease* (GERD) when the stomach acid that refluxes back into the esophagus causes inflammation and pain. The problem for parents and doctors is trying to figure out when a baby is in pain.

The list of symptoms that distinguish GER from GERD is long and largely subjective. For example, babies are said to be in pain if they arch their back or neck during feeding. (Many babies do this anyway.) Vomiting is said to be more forceful with GERD than with GER. (To most parents, all vomiting appears to be forceful.) Babies are said to be more irritable with GERD than with GER. (All crying babies appear to be very irritable, especially after midnight.) But the most objective way to distinguish GER from GERD is that the pain of GERD is so bad that babies eat less, causing them to fall off their growth curve—that is, they fail to thrive.

Two different types of medicine are available to alleviate the symptoms of GERD. Their differences center on how they work to decrease stomach acid.

The cells in the stomach that produce acid are called parietal cells. Several different chemicals produced in the body stimulate parietal cells. One is histamine. Medicines that block the attachment of histamine to parietal cells, thereby decreasing acid production, are called histamine receptor antagonists. Zantac is a histamine receptor antagonist.

The other type of reflux medicine targets the central mechanism of acid production. Parietal cells make acid by exchanging two protons: hydrogen (H^+) and potassium (K^+). Potassium is pumped into the cell at the same time that hydrogen is pumped out. Once in the stomach, hydrogen pairs with chloride to form hydrochloric acid, one of the most powerful acids in nature. Medicines that inhibit this proton pump are called proton pump inhibitors (PPIs). Prilosec and Prevacid are PPIs.

Two facts about using reflux medicines to treat fussy babies will surprise you. First, they don't work. Second, they're dangerous.

Reflux Medicines Don't Work to Treat Fussy Babies

In 2003, researchers from the University of Australia divided thirty irritable infants with reflux into two groups. One group was given reflux medicines every day for four weeks; the other, a placebo. The amount of irritability between the two groups was indistinguishable.

In 2009, researchers from Australia and Sweden divided thirty-eight infants with reflux into two groups. One group received reflux medicines every day for seven days; the other, a placebo. Although the reflux medicine group made less stomach acid than the placebo group, the amount of stomach contents regurgitated was the same.

Also in 2009, researchers from the United States and Canada divided 162 infants with reflux into two groups. One group received reflux medicines every day for four weeks; the other, a placebo. No differences in the incidence of feeding-related crying were observed. There was, however, one disturbing finding. Babies treated with reflux medicines were more likely to suffer pneumonia than those given the placebo. (The reason for this rather surprising finding will be explained later.)

In 2010, researchers from the United States and Poland divided 106 infants with reflux into two groups. One group received a reflux medicine every day for four weeks; the other, a placebo. The authors found no differences in symptoms between the two groups.

THE problem with reflux medicines, however, isn't only that they don't work. It's that they're dangerous.

Reflux Medicines Are Dangerous

Both histamine receptor antagonists and PPIs have side effects, some of which are quite serious. To understand why, consider the following: The stomach contains about one billion parietal cells. The proton pump used by these cells to make acid requires more energy than any other cellular process, including the conversion of food to energy, the generation of immune responses, and the rhythmic beating of the heart. To generate this energy, humans pay a high caloric cost. It is, therefore, reasonable to assume that the production of stomach acid is an important adaptation to our environment—a process necessary for our survival.

Why is stomach acid so important? Answering this question requires understanding where humans fit among other species, some of which make far more stomach acid than we do. One such species is the turkey vulture, nature's premier scavenger. Turkey vultures eat carrion, the decaying meat of dead animals. Carrion is riddled with bacteria that can cause disease. The purpose of stomach acid is to kill these bacteria before they enter the intestine. Red-tailed hawks also generate more acid in their stomachs than humans. Unlike turkey vultures, however, red-tailed hawks are occasional scavengers, eating carrion only when convenient. Wild chickens, which are carnivores, are strict meat eaters. They, too, make more stomach acid than humans. Humans, however, are omnivores, meaning that we eat meat as well as plants. Because we eat meat from related species (i.e., other mammals), and because these related species carry bacteria that can hurt us, we are fairly high on the scale of acid producers. Animals that produce less stomach acid than humans include elephants and colobus monkeys, both of which are herbivores. Because plants and animals are far apart on the phylogenetic tree, plants are much less likely to carry bacteria that hurt animals. For this reason, herbivores don't put much energy into making stomach acid. In other words, the best strategy to avoid potentially harmful bacteria is to consume food that is distant to your species. Birds that consume insects are at lower risk of bacterial infections than birds that consume other birds.

It shouldn't be surprising, then, that infants treated with medicines that reduce acid production are at greater risk of bacterial infections of the intestine. Infants are also at greater risk of acquiring bacteria that eventually travel to the lungs and cause pneumonia. Worse, low-birth-weight infants treated with reflux medicines are at risk of a severe and occasionally fatal disease called necrotizing enterocolitis, caused by a breakdown of the intestinal surface. Most recently, a study of more than 850,000 babies found that those given reflux medicines had an increased risk of bone fractures.

The presence of stomach acid also determines the quantity and type of the hundreds of trillions of bacteria that live at the intestinal surface (the microbiome). These bacteria can determine the likelihood of developing obesity, allergies, diabetes, and a host of other problems. Indeed, a study of 800,000 children of military personnel performed between 2001 and 2013 found that those who had received reflux medicines were more likely to develop food allergies, medication allergies, acute allergic reactions (anaphylaxis), allergic rhinitis (runny nose), and asthma.

The good news is that some children who are falling off their growth curves because of GERD can be treated successfully without reflux medicines. Although these strategies don't always work, they're worth a try. For example:

- Cow's milk allergy can mimic the symptoms of GERD. If that's the problem, GERD should resolve within two to four weeks of eliminating cow's milk from the diet.

- Women who are breastfeeding babies with GERD should eliminate milk, eggs, coffee, and spicy foods.

- For children who are formula fed, parents should reduce the volume and increase the frequency of feedings. Also, thickening formula with one tablespoon of rice cereal per ounce can help.

- Positioning therapy can also be curative. Just place the child in a prone (on the stomach) or upright position after feeding. Or put a wedge pillow under the sheet to elevate the head and upper body.

Although these dietary and positional therapies might not work, they're worth a try. And they're far better than exposing babies to a medicine that doesn't work and that might be harmful.

Babies cry when they're overtired. They cry when they're overstimulated. And when they cry, they clench their fists and arch their backs and spit up. It's easy to understand how parents could assume that their babies are in pain. But the simple truth is that all this crying, at least most of the time, has nothing to do with acid reflux. Sadly, magic pills to stop crying don't exist.

In 2019, the official medical journal of the American Academy of Pediatrics stated that "multiple studies have failed to show reduction in clinical symptoms of crying, coughing, and back arching with [reflux medicines] compared with placebos. These studies also did not demonstrate improvement in regurgitation." Nonetheless, one out of every ten babies in the United States is treated with reflux medicines.

PART IV

Cancer Screening

11

Prostate Cancer Screening Programs Do More Harm Than Good

Cancer prevention programs are best understood by imagining a barn containing three animals: a bird, a turtle, and a rabbit. When you open the barn door, the bird flies out immediately. It's just too quick. The bird represents a type of cancer that is so virulent, so overwhelming, that even early detection won't save lives. The turtle, however, is too slow to escape. Even if you held the barn door open for days, the turtle wouldn't get out. The turtle represents a type of cancer that is so slow growing that people die *with* it, not *from* it. Early detection of turtles is unnecessary and often harmful. Next up is the rabbit. The rabbit is fast but not as fast as the bird. It's possible to close the barn door before the rabbit escapes. The rabbit represents a type of cancer whose early detection will save lives.

Several cancer screening programs clearly make a difference. Pap smears prevent fatal cervical cancers, and colonoscopies prevent fatal colon cancers. Unfortunately, not all cancer screening programs are as effective. Prostate, thyroid, and breast cancer screening programs, for example, might be doing more harm than good. Atul Gawande, a renowned cancer surgeon at Brigham and Women's Hospital, in Boston, summed it up best: "We now have a vast and costly healthcare industry devoted to finding and responding to turtles."

PROSTATE cancer is the most common cancer in men, second only to lung cancer as a cause of death. In 2017, 161,000 men in the United States were diagnosed with prostate cancer, and 27,000 died from the disease. Indeed, one out of every seven men will be diagnosed with prostate cancer in his lifetime. Given the magnitude of this problem, an effective screening program would be of enormous benefit.

Cancer screening programs differ in what they measure, how they measure it, and when they measure it. Pap smears, for example, detect precancerous cells in the cervix. Colonoscopies detect precancerous polyps in the intestine. Prostate screening programs, however, don't detect abnormal cells, suspicious polyps, or worrisome lumps. Instead, the initial screening test for prostate cancer is a blood test that detects an enzyme made in the prostate called prostate-specific antigen (or PSA).

Normally, the testes produce sperm in a thick mucus called a coagulum. PSA liquefies this coagulum so that sperm can swim freely. PSA also dissolves cervical mucus so that sperm can enter the uterus on their way to fertilizing an egg. The first practical use of PSA was as a forensic marker in rape cases. (Rapists who don't make sperm in their testes still make PSA in their prostate, even after a vasectomy.) The researcher who discovered PSA, Richard Ablin, never imagined that his test would be used to predict whether someone was at increased risk for prostate cancer, for a few reasons:

- PSA is also detected in breast milk and amniotic fluid. So, it's not really prostate specific.

- Men with prostate cancer can have very low levels of PSA, and men without prostate cancer can have very high levels. Also, other common problems, such as enlargement or infection of the prostate, can elevate PSA levels. So, it's not really cancer specific.

Nonetheless, in 1994, the Food and Drug Administration approved PSA levels as a screening test for prostate cancer.

Perhaps no one popularized PSA tests more than Bob Dole, a Republican senator from Kansas. In 1992, during the Republican National Convention in Houston, Dole, a prostate cancer survivor, urged men to visit a designated tent for a free PSA screening. Later, Dole spearheaded similar booths at the Kansas State Fair, the Chicago Mercantile Exchange, and, in 1996, the Republican National Convention in San Diego, where he was his party's nominee. During the 1996 convention, about twenty thousand men got free PSA screenings.

Once PSA levels were approved for screening, advocacy groups, urologists, and hospitals embraced them, assuming the test would save lives. The rules were clear. If the PSA level was less than 1 nanogram, no further testing was done. (A nanogram is one-billionth of a gram.) If the PSA level was between 1 and 3 nanograms, testing was repeated every two to four years. If the PSA level was greater than 3 nanograms, urologists biopsied the prostate. If the prostate biopsy showed no evidence of cancer, then PSA levels were obtained every one or two years. If PSA levels had returned to normal, biopsies were discontinued. If PSA levels remained high, repeat biopsies were performed. If a biopsy detected prostate cancer, the prostate was either removed surgically or irradiated. Physicians have embraced this strategy for reducing deaths from prostate cancer for more than twenty years. Four studies have examined its impact.

Prostate Screening Programs Don't Save Lives

The Prostate, Lung, Colorectal, and Ovarian Cancer Screening (PLCO) Trial assigned 76,000 men ages fifty-five to seventy-four to annual PSA screenings or usual care. Fifteen years later, the number of deaths from prostate cancer was the same in both groups. PSA screenings, biopsies, surgeries, and radiation therapy had done nothing to prolong lives.

The European Randomized Study of Screening for Prostate Cancer (ERSPC) assigned 160,000 men ages fifty-five to sixty-nine to PSA screening or usual care. Unlike the PLCO study, ERSPC found that prostate screening programs saved lives, but the number of lives saved was far from dramatic: 781 men had to be screened to prevent one man from dying from prostate cancer over thirteen years. There was, however, a physical and psychological price to be paid for preventing this death. We'll get to that.

The Cluster Randomized Trial of PSA Testing for Prostate Cancer (CAP) assigned 415,000 men to receive a PSA test or not. After ten years, similar to the PLCO trial, the number of deaths from prostate cancer was the same in both groups.

The results from the Prostate Testing for Cancer and Treatment (ProtecT) study were probably the most sobering. Between 1999 and 2009, 82,000 men ages fifty to sixty-nine received a PSA test; 2,664 were diagnosed with prostate cancer by biopsy. Of those diagnosed with prostate cancer, 1,643 were randomized to either a complete removal of the prostate (prostatectomy), radiation therapy, or active monitoring. Seventeen men in this study died from prostate cancer. Remarkably, no differences were detected among the three groups. Men were just as likely to die from prostate cancer if they had surgery or radiation as they were if nothing was done.

Why haven't prostate screening programs saved lives? Several explanations have been offered: (1) most of the prostate tumors detected were so slow growing (i.e., turtles) that men died *with* these cancers, not *from* them; (2) some of these cancers were so virulent (i.e., birds) that early detection didn't matter; (3) in some cases, the immune system recognized an early cancer and eliminated it; and (4) in others, the cancer outgrew its blood supply and regressed on its own. In any case, men screened for prostate cancer haven't lived longer than those who weren't screened.

When deciding whether to enter a prostate screening program, men should consider the following facts:

- Most men die *with* prostate cancer, not *from* it. Autopsies of men ages fifty to fifty-nine who had died from other causes found that 20 percent had prostate cancer. In men ages seventy to seventy-nine who had died from other causes, 33 percent had prostate cancer at autopsy.
- Most men with high PSA levels are biopsied. Side effects from prostate biopsies include pain, fever, and blood in sperm (hematospermia). Further, one out of every one hundred prostate biopsies causes complications so severe that patients need to be hospitalized. Prostate biopsies are also a rare cause of bacterial infections of the bloodstream (sepsis), which can have a death rate as high as 30 percent.
- For those diagnosed with prostate cancer, surgical removal of the prostate will cause one in eight men to suffer long-term urinary incontinence, forcing them to wear a pad for the rest of their lives. Further, about one in three will experience long-term erectile dysfunction. Worse, three out of every one thousand men die either during or soon after a prostatectomy. Side effects from radiation therapy to the prostate are similarly grim; one in two men experience long-term erectile dysfunction, and one in six experience uncomfortable bowel symptoms including urgency and incontinence.

HERE'S a simpler way to look at this:

For every 1,000 men screened, 240 will have a high PSA.

Among the 240 with a high PSA, 100 will have a biopsy that reveals prostate cancer.

Among the 100 with a positive biopsy, 80 will choose either prostatectomy or radiation therapy.

Among the 80 who chose either prostatectomy or radiation therapy, 50 will suffer a complication from the treatments.

Among the 50 who suffer complications from the treatments, 15 will suffer long-term erectile dysfunction; 15 will suffer from long-term urinary incontinence; 5 will die from prostate cancer even after radical therapy; and all who had a positive PSA test will suffer anxiety and sleepless nights.

Summing up, among the 1,000 men initially screened, 1 might have his life saved, even though 100 will be diagnosed with cancer. The other 99 will have undergone surgery or radiation, and all their attendant complications, without benefit.

PSA screening and prostate biopsies have caused an emotional, expensive, physically harmful pseudo-epidemic of prostate cancer yet haven't allowed men to live longer.

As a consequence of these findings, the American Academy of Family Physicians and the Canadian Task Force on Preventive Health Care recommend against prostate screening programs. Other advisory committees have taken a more nuanced position. Based largely on the results of ERSPC, which found that one life could be saved for every 781 men screened, the U.S. Preventive Services Task Force, the American College of Physicians, and the American Urological Association recommend that doctors discuss the significant risks and possible benefits of prostate screening programs with patients before ordering a PSA level screening.

During the past two decades, because of the disappointing results of these studies, the number of men entering prostate cancer screening programs has been cut in half. Nonetheless, thousands of men continue to subject themselves to these programs every year.

IN 2010, Richard Ablin, the researcher who discovered PSA, penned an op-ed for the *New York Times*. Appalled by how his PSA test had been misused, he wrote, "I never dreamed that my discovery four decades ago would lead to such a profit-driven public health disaster."

Given that current prostate cancer screening programs aren't the answer, isn't there something that can be done to prevent this common, often fatal cancer? It's possible, but it won't be easy. Researchers must first define either biological or genetic markers in prostate cells obtained during biopsy that clearly distinguish aggressive from nonaggressive cancers. They must also develop better immunological or biological markers for response to therapies. And perhaps most difficult of all, they must develop better imaging techniques. With these tools in hand, doctors will finally be able to say, "This is a type of prostate cancer that will kill you" and "this is a type that won't." As it stands, PSA screenings, biopsies, surgery, and radiation have done little to nothing to help clinicians defeat this cancer and have left many men impotent and incontinent.

12

Thyroid Cancer Screening Programs Don't Save Lives

We're in the midst of an epidemic. Between 1975 and 2002, the incidence of thyroid cancer in the United States doubled. By 2009, it had tripled—a growth rate faster than that of any other cancer. By 2013, more than 630,000 people in the United States were living with thyroid cancer. This year, another 60,000 Americans will be diagnosed with the disease, and 2,000 will die from it. Women will suffer the most, outnumbering men by three to one.

The United States isn't alone. During the past three decades, in France, Italy, Croatia, Israel, China, Australia, Switzerland, Canada, and the Czech Republic, the rate of thyroid cancer has also more than doubled.

No country, however, has suffered this epidemic of thyroid cancer more than South Korea. In 1999, the South Korean government launched a nationwide cancer screening program. Every adult was screened for breast, cervical, colon, stomach, and liver cancer—all free of charge. Thyroid cancer screening was optional, but it was still inexpensive: citizens could be screened for it for the equivalent of thirty to fifty dollars. Most chose this option. As a consequence, between 1999 and 2011, the incidence of thyroid cancer in South Korea *increased* fifteenfold! By 2012, more than forty thousand South Koreans were being diagnosed with thyroid cancer every year. South Korea now had the highest incidence of thyroid cancer in the world. It was, as one researcher noted, a "tsunami of thyroid cancer."

Perhaps most surprising, all these people newly diagnosed with thyroid cancer didn't have symptoms at the time of diagnosis (e.g., hoarseness, difficulty swallowing, painless swelling in the neck, difficulty breathing, or a change in the tone or quality of their voice). These patients were lucky, it seemed, that they had been diagnosed so early, well before their cancers could kill them.

What accounts for this dramatic increase in thyroid cancers across the globe?

One possibility is exposure to radiation, to which the thyroid gland, located in the front of the neck, is particularly susceptible. For example, in August 1945, at the end of World War II, atomic bombs were dropped on Hiroshima and Nagasaki. In addition to at least 140,000 people dying immediately, thousands more died from a variety of cancers, including thyroid cancer, within a few years.

Nuclear bomb testing has also caused unwanted radiation exposure. Between June 1946 and November 1962, the United States performed above- and belowground tests of nuclear weapons in the Marshall Islands, on Christmas Island, in the Johnston Atoll in the Pacific Ocean, and over the South Atlantic Ocean.

More radiation exposures followed. In April 1986, in the wake of an explosion at the Chernobyl nuclear reactor in Ukraine, thousands of people died from radiation-induced cancers. Many of them were children who had died of thyroid cancer.

Radiation exposure, however, cannot explain the current epidemic. The deployment of atomic bombs during World War II, the testing of nuclear bombs in the early 1960s, and the mismanagement of nuclear power plants into the 1980s all occurred well before the current increase in thyroid cancer.

A second possibility, then, is the purposeful and ill-conceived use of radiation to "treat" common disorders such as acne, tonsillitis, ringworm (a fungal infection), and mastitis (inflammation of the breasts). Radiation has also been used to shrink enlarged adenoids and thymus glands. Fortunately, this dangerous practice ended in the 1950s—again, well before the current epidemic of thyroid cancer.

A third possibility is the influence of advocacy groups. Some people who have survived thyroid cancer, or who had relatives or friends who died from the disease, have launched campaigns to increase awareness. One campaign, sponsored by the Light of Life Foundation, is called "Check Your Neck." Advertisements featured Brooke Shields and Cindy Crawford over the tagline "It's not just looks that can kill. Have your neck checked for thyroid cancer." But again, where radiation exposures occurred too early to explain the increase in thyroid cancer, advocacy efforts started too late. "Check Your Neck," for example, didn't launch until 2014, well after the epidemic of thyroid cancer had begun.

One explanation, however, fits perfectly. Several decades ago, two medical inventions turned the incidence of thyroid cancer on its head.

In the 1980s, biomedical engineers invented ultrasound machines, which were quickly adopted for widespread use. Capable of detecting thyroid lumps as small as 0.2 centimeters (about 0.08 inches), ultrasounds revolutionized the field of cancer detection. The premise of ultrasounds is simple. Using a handheld probe, the ultrasound machine emits sound waves higher than those audible to humans. These sound waves bounce off the thyroid gland (or any other organ in the body) and are then transferred back to the operator, where they are displayed as an image. Different structures with different densities absorb or reflect sound waves differently.

Ultrasounds have several advantages over previously available imaging techniques: (1) they provide images immediately; (2) they're portable and can be used in any doctor's office; (3) they're cheap; and (4) they avoid potentially harmful radiation from nuclear scans, which previously had been used to detect thyroid cancers. Best of all, ultrasounds are ten times more sensitive than a doctor's physical exam at detecting small lumps in the thyroid.

The second technical advance was called "fine needle biopsy." No longer did patients have to go to a surgeon for biopsies of suspicious lumps in their necks. Like ultrasounds, fine needle biopsies could be performed in a doctor's office without the need for anesthesia. By the late 1990s, with the ready availability of ultrasounds and fine needle biopsies, the diagnosis of thyroid cancer had become an outpatient procedure, free from the need for hospitals, surgeons, or radiologists. And with it came a "tsunami" of cancer diagnoses.

Now, with the detection of so many early, small, symptom-free thyroid cancers, the only thing left to do was to treat them, which meant total removal of the thyroid (thyroidectomy) followed, in some cases, by ingestion of radioactive iodine to eliminate any remaining cancer cells. During the past two decades, 150,000 women and 50,000 men have undergone thyroidectomies, a procedure that is far from harmless. For example, the laryngeal nerve, which innervates the vocal cords, is located right next to the thyroid gland. About two out of every one hundred thyroidectomies cause permanent vocal cord paralysis, symptoms of which include hoarseness, noisy breathing, the inability to speak loudly, a breathy quality of the voice, loss of vocal pitch, the need to breathe frequently while speaking, and choking or coughing while swallowing food, drink, or saliva.

Another problem centers on a structure next to the thyroid gland called the parathyroid gland, which controls calcium balance. About six out of every one hundred thyroidectomies result in the loss of this gland and, as a result, abnormal levels of calcium in the bloodstream and bones.

Finally, radioactive iodine, which is occasionally ingested by patients following a thyroidectomy, increases the risk of leukemia.

But the worst side effect of total thyroidectomy, and one that is experienced by virtually everyone who has had this procedure, is the need to replace the hormone produced by the thyroid, which is critical to the body's metabolism. It's really hard to get the dose right. Too much thyroid hormone can cause sweating, fatigue, heat intolerance, restlessness, mood swings, excessive hunger, nervousness, panic attacks, abnormal heart rates, palpitations, insomnia, hyperactivity, irritability, abnormal protrusion of the eyes, weight loss, diarrhea, hair loss, muscle weakness, tremor, and warm skin. Too little thyroid hormone can cause lethargy, brittle nails, dry hair, constipation, sensitivity to cold, sexual dysfunction, slow heart rate, sluggishness, weight gain, and irregular uterine bleeding.

Thyroidectomies have come with an enormous price, but at least patients are still alive. At least they no longer have to die from undiagnosed and untreated thyroid cancers. Unfortunately, it hasn't worked out that way. Most thyroid cancer sufferers would be surprised to learn that all these ultrasounds and fine needle biopsies and thyroidectomies and radioactive iodine treatments and hormone replacement therapies haven't saved lives. *Indeed, the incidence of death from thyroid cancer hasn't budged for the last twenty years*. This has been true not only for the United States, but also for South Korea, Switzerland, and every other country that has suffered both a tsunami of thyroid cancer and a tsunami of thyroidectomies and their harms. "An epidemic of real disease would be expected to produce a dramatic rise in the number of deaths from the disease," said Gilbert Welch, a professor of medicine at Dartmouth. "Instead, we see an epidemic of diagnosis . . . an epidemic of our own creating." Perhaps most concerning, the twenty-year survival rate for people with thyroid cancers who did or didn't have thyroidectomies is statistically indistinguishable.

Why haven't all these early diagnoses and early treatments of thyroid cancer saved lives? The answer can be found in the type of thyroid cancer that is being detected, something pathologists call papillary cancer. More than 95 percent of the cancers detected by screenings and biopsies of patients without symptoms have been of this type. What we have now learned—but far, far too late—is that papillary cancers probably shouldn't be called cancers at all.

The first evidence that papillary cancers weren't what they were claimed to be came from autopsy studies. In 1985, researchers in Finland examined the thyroids of 101 people who had died for reasons other than thyroid cancer. They detected papillary cancers in thirty-six thyroid glands. These same researchers reasoned that had they made their sections of the thyroids finer, they would have detected papillary cancers in 100 percent of the specimens.

Japanese researchers later extended the Finnish study, finding many papillary cancers in autopsy specimens in young adults who also had died from something other than thyroid cancer. John C. Morris, president of the American Thyroid Association and a professor of medicine at the Mayo Clinic, stated the obvious: "If it's not cancer," he said, "let's not call it cancer."

In the barn analogy described in the chapter on prostate cancer screening, birds are cancers that are so virulent they kill you no matter how early they are detected; turtles are cancers that are so slow growing that you are more likely to die with them than from them; and rabbits are cancers that, if detected early, can be survived. Small papillary cancers of the thyroid aren't birds, rabbits, or turtles. They aren't cancers at all and, according to the U.S. Preventive Services Task Force, should now be "classified as a normal finding." This isn't to say that people don't die from thyroid cancer. They do. Every year, about two thousand people in the United States die from this disease. But these fatal thyroid cancers aren't papillary cancers; they're anaplastic cancers, and they're almost always birds.

Because thyroid screening programs have caused an avalanche of unnecessary surgeries and consequent harms, advisory bodies such as the U.S. Preventive Services Task Force, the American Academy of Family Physicians, the American Thyroid Association, the American College of Endocrinology, the American Association of Clinical Endocrinologists, and the Canadian Task Force on Preventive Health Care all recommend against thyroid screening programs in patients without symptoms of thyroid cancer.

Nonetheless, despite these unambiguous recommendations from almost every advisory body in the world, tens of thousands of Americans this year will have unnecessary ultrasounds, unnecessary biopsies, and unnecessary surgeries causing vocal cord damage, difficulties with calcium regulation, and problems with thyroid replacement hormones. The assault on American thyroids continues unabated. Why is this still happening?

The first explanation, offered by Dr. Barnett Kramer, director of the Division of Cancer Prevention at the National Cancer Institute, is probably the most obvious. Recommendations *not* to screen for potentially fatal cancers are counterintuitive. "The message for so long is that early detection is always good for you," said Kramer. But the false promise and harms of thyroid screening, he argued, "should acutely raise awareness of the consequences of acting on the intuition that *all* screening must be of benefit."

The second explanation is the easiest to understand. Many patients are diagnosed with thyroid cancer incidentally while looking for causes of thinning hair or weight gain or other symptoms that are generally unrelated to cancer but could be related to an overactive or underactive thyroid. "Once we have made a diagnosis of cancer it is difficult to say, 'Don't do anything,'" said Dr. Ashok R. Shaha, a thyroid cancer surgeon at Memorial Sloan Kettering Cancer Center in New York City. "In the U.S., we have a fear that if we miss a cancer the patient will sue."

The third explanation, and the most nefarious, is that doctors get paid for what they do, not for what they don't do. Ultrasounds, biopsies, surgeries, and drugs cost money. Watchful waiting, on the other hand, is free. Dr. Michael Tuttle, an endocrinologist at Memorial Sloan Kettering and a member of the American Thyroid Association's guidelines committee, disagreed. "You don't make a lot of money doing ultrasounds and biopsies," he said. "But there is no question that you get paid if you do that, and surgeons are going to be paid to do the surgery." Nonetheless, Tuttle argued, "it just doesn't feel to me like monetary incentives are what is driving it. What's driving it is the desire to make a diagnosis of cancer and treat cancer. When I talk to doctors who are doing early diagnosis and treatment, they truly believe in their heart of hearts that they did a very good thing by finding a [small] papillary thyroid cancer. I find that when I give lectures and talk about these issues, the people who give me grief are not worried about incentives or cost or money. They think that by not doing something, we are preventing them from curing some small cancer."

What will it take to finally put an end to unnecessary thyroid cancer screenings—or, at the very least, to put an end to unnecessary surgeries? One hopeful sign is that some cancer centers are electing for watchful waiting. "I saw someone today whose hair was getting thin and her doctor decided to do an ultrasound to check the thyroid," recalled Tuttle. "Of course, they found a [small] papillary thyroid cancer that we are now going to follow with observation." It's a lot to ask. Once people believe that they have cancer, it's hard to get them to live with the fact that the cancer is still in them. They want it taken out. The notion that they have to wait feels dangerous. The only real solution is for pathologists to find another name for papillary cancers, something far less frightening.

Perhaps no group of physicians has fought harder to end the war on the thyroid than those who have witnessed the most damage. In March 2014, eight South Korean doctors formed the Physician Coalition for Prevention of Overdiagnosis of Thyroid Cancer. Their first act was to write an open letter to the public highlighting the problem. The response was immediate. Television stations ran hourlong investigative reports, and newspapers ran articles with headlines like "What Caused Jump in Thyroid Cancer Cases?" and "Is Thyroid Cancer Overdiagnosed?" This small group of South Korean physicians had dared to take on a lucrative industry. Thyroid screening programs and surgeries had become big business in South Korea. Hospitals had expanded their thyroid clinics and hired more surgeons. And companies that made devices to assist robotic surgeries were reaping enormous profits. In response to the coalition's open letter, the Korean Thyroid Association, a professional society of endocrinologists and thyroid surgeons, pushed back, writing that screening and treatment shouldn't be curtailed because they are a "basic human right." Despite this opposition, the coalition has had an impact. By 2015, the number of thyroidectomies in South Korea had declined from 43,000 to 28,000—a 35 percent decrease. Also, in 2015, the Korean Guideline for Thyroid Cancer Screening was finally modified, stating that "thyroid ultrasonography is not routinely recommended for healthy subjects."

Change is possible, but first, doctors who routinely perform unnecessary neck ultrasounds and thyroid biopsies need to become better advocates for their patients, and patients need to become better advocates for themselves.

13

Breast Cancer Screening Programs Aren't Exactly as Advertised

Every year in the United States, hundreds of thousands of women are diagnosed with an early stage of breast cancer. Almost all believe that this early diagnosis has saved their lives. As it turns out, only one in four is right. The other three have to suffer the complications of surgery, radiation therapy, hormone therapy, and chemotherapy with no benefit. Most of these early breast cancers are actually *pre-cancerous* tumors that may or may not progress to true cancers. How can we tell who benefits from these early diagnoses and who doesn't?

Breast cancer is the most common cancer in women worldwide. Every year, more than 1.5 million women are diagnosed with breast cancer and 500,000 die from the disease. Breast cancer is second only to lung cancer as the most common cause of cancer death in women.

In the United States, more than 200,000 women are diagnosed with breast cancer every year, and 40,000 die from the disease. Indeed, one in eight women will be diagnosed with breast cancer.

When aggressive, the cancer spreads to the liver, brain, bone, and lungs. For this reason, it's important to diagnose it early.

IN 1895, Wilhelm Röntgen detected and produced X-rays, electromagnetic radiation of high energy and short wavelengths that pass through anything opaque to light, including human skin. More than fifty years later, Robert Egan, while working at the University of Texas MD Anderson Cancer Center, modified Röntgen's discovery. Using much lower levels of radiation, Egan was able to detect abnormal lumps as small as a fraction of an inch, tumors that could never have been detected by physical examination. In 1958, Egan published a paper about it, and in 1964 he published a book, *Mammography*. The name stuck.

The first person to champion mammography was a radiologist named Philip Strax. In the early 1960s, in collaboration with Sam Shapiro, a statistician, and Louis Venet, a surgeon, Strax performed a study proving that mammography could detect breast cancers undetectable by physical exam. Mammography, argued Strax, could save lives. Then he proved it.

In December 1963, working with the Health Insurance Plan of Greater New York, Strax, Shapiro, and Venet divided 62,000 women between ages forty and sixty-four into two groups. One group received yearly mammograms; the other didn't. Six years later, the researchers found that thirty-one women in the mammography group had died of breast cancer and fifty-two in the no-mammography group. In a paper published in the *Journal of the American Medical Association*, the researchers concluded, "The present communication offers the first hard evidence on the impact that screening has on mortality and the results are encouraging."

This was a breakthrough study. Between 1935 and 1967, the death rates for women from all causes had decreased by 42 percent. The death rate from breast cancer, however, hadn't budged. The Strax study offered the first ray of hope that things could change. In

1988, Strax and Shapiro shared the prestigious Kettering Prize for outstanding contributions in the field of cancer.

More studies followed.

In 1985, researchers in Sweden divided 135,000 women age forty or older into two groups. One group was offered screening mammograms every two to three years; the other wasn't. Subjects were followed for seven years. At the end of the study, those receiving mammograms who were age fifty or older had a 31 percent decrease in death from breast cancer.

In 1993, researchers from medical schools in New Hampshire, North Carolina, and Maryland reviewed the results of eight studies that involved 500,000 participants. They, too, found that screening mammography had decreased the incidence of fatal breast cancer by about a third in those participants over age fifty. The authors enthused, "Randomized trials have provided stronger scientific evidence regarding the effectiveness of screening for breast cancer *than for any other cancer*" (emphasis mine).

The results were clear. Mammography (a procedure that took only twenty minutes) could save lives. By the end of the 1980s, about 70 percent of U.S. women were getting them. The incidence of deaths from breast cancer steadily decreased.

Then the picture darkened and blurred. As a consequence, some countries no longer recommend mammograms for their citizens. What happened?

IN 2000, researchers from the Nordic Cochrane Centre in Denmark reviewed the results of eight studies of women who did or didn't get mammograms, some of which were performed better than others. They concluded, "The two adequately randomized trials found no effect on breast cancer mortality or on total mortality."

In 2013, these same researchers reviewed the results of several more studies that followed 600,000 women ages thirty-nine to seventy-four who did or didn't get mammograms and found that "The trials with adequate randomization did not find an effect of screening on total cancer mortality, including breast cancer, after ten years."

In 2014, Canadian researchers completed a twenty-five-year study in which they'd divided 90,000 women between ages forty and fifty-nine into two groups. One group received screening mammograms every year; the other didn't. At the end of the study, 3,250 women in the mammography arm were diagnosed with breast cancer, and 500 died from the disease. In women who had never received mammograms, 3,133 were diagnosed with breast cancer, and 505 died from the disease—a statistically insignificant difference. And it didn't matter which age groups were examined. Differences weren't observed in women between ages forty and forty-nine or ages fifty and fifty-nine. No one, apparently, had benefited. The authors concluded that the "value of mammography screening should be reassessed." Then they gave an important clue as to what was happening: "Annual mammography in women aged 40 to 59 does not reduce mortality from breast cancer beyond that of physical examination or usual care *when adjuvant [additional] therapy for breast cancer is freely available*" (emphasis mine). It was this last phrase that hinted at why researchers were now having so much trouble proving that mammography saved lives.

WHAT did the authors of the 2014 Canadian study mean when they wrote "adjuvant therapy for breast cancer"? During the past twenty years, advances in the treatment of breast cancer have been remarkable, owing mainly to our ability to define various proteins on the surface of breast cancer cells. Now breast cancers can be divided into three types. Some breast cancer cells bear receptors on their surface for the hormones estrogen or progesterone. Others express a protein called human epidermal growth factor 2 (ERBB2, formerly Her2). Yet another group of cancers (so-called triple-negative breast cancer) doesn't express any of these proteins. Different breast cancer types require different therapies. Today, in contrast to the time when mammography was first studied, cancer-targeted antibodies, hormone therapies, and chemotherapeutic drugs specific for different types of breast cancer are available. As a consequence, the five-year survival rates for early breast cancers that express either estrogen or progesterone or ERBB2 are 94 to 99 percent! For triple-negative cancers, the five-year survival rate is 85 percent. For this reason, even women whose breast cancers aren't detected early have an excellent prognosis.

The impact of breast cancer treatments becomes clearer when examining mortality rates in different parts of the world. In resource-rich countries, such as the United States, the five-year survival rate for breast cancer exceeds 80 percent. In resource-poor countries, where therapies are less available, the five-year survival rates are between 10 and 40 percent.

At this point, one could argue that at worst, mammography provides a marginal benefit in preventing deaths from breast cancer. But there's a bigger problem. Early screening with mammograms has caused many women to suffer needlessly. No study showed just how damaging mammography can be more than one published in the *New England Journal of Medicine* several years ago.

In 2012, Archie Bleyer, from the Oregon Health and Science University, in Portland, and Gilbert Welch, from Dartmouth's Medical School, published a paper titled "Effect of Three Decades of Screening Mammography on Breast Cancer Incidence." Bleyer and Welch found that since the inception of breast cancer screening programs in the United States, the number of cases of early breast cancer had doubled, from 112 to 234 cases per 100,000 women screened. Therefore, every year, 122 more women per 100,000 were being diagnosed with breast cancer. On the surface, these numbers were encouraging. Now more than 100 women per 100,000 no longer had to die from breast cancer. But that's not what these researchers found. Rather, they found that the rate of late-stage breast cancer had decreased by only 8 percent, from 102 to 94 cases per 100,000. This meant that only 8 of those 122 women who had been diagnosed with early-stage breast cancer had been prevented from progressing to late-stage disease. In other words, only 8 women per 100,000 had had their lives saved by early screening. But what about the other 114 women? Why hadn't they had their lives saved by mammography? One possibility is that these early cancers had either receded on their own or were so slow growing that the women would have invariably died from something else; meaning they died *with* their breast cancer, not *from* it. Another possibility is that the cancer was so easily cured that it didn't matter when it was detected—a consequence of better treatments. A final possibility is that the cancer was so virulent that it didn't matter when it was detected; these women would have died from their breast cancer independent of whether the diagnosis had been made early or late.

One thing, however, was clear from Bleyer and Welch's study. At least 114 women per 100,000 had undergone a combination of surgery, radiation, chemotherapy, or hormonal therapy without benefit. Bleyer and Welch estimated that during the previous thirty years, about 1.3 million women in the United States had been treated aggressively for breast cancers that never would have killed them, and that about 70,000 women were continuing to be overdiagnosed and overtreated every year. The authors concluded that "Despite substantial increases in the number of cases of early stage breast cancer detected, screening mammography has only marginally reduced the rate at which women present [to their doctor] with advanced breast cancer. Although it is not certain which women have been affected, the imbalance suggests that there is substantial overdiagnosis, accounting for nearly a third of all newly diagnosed breast cancers, and that *screening is having, at best, only a small effect on the rate of death from breast cancer*" (emphasis mine).

Fewer women today are dying from breast cancer. The Bleyer and Welch study had shown why. "We are left to conclude, as others have, that the good news in breast cancer—decreasing mortality—must largely be the result of improved treatment, not screening," they wrote. The bad news was that the physical and emotional cost of early detection with mammography was high. Surgeries are disfiguring. Radiation therapy increases the risk of heart disease and lung cancer. Hormone therapies cause a variety of unpleasant side effects. And chemotherapies cause anemia, nausea, vomiting, hair loss, bone thinning, heart toxicity, and immune dysfunction leading to severe and occasionally fatal infections.

Bleyer and Welch's findings were reproduced by others. In 2012, researchers from Harvard Medical School evaluated a nationwide screening program in Norwegian women ages fifty to sixty-nine. They found that for every 2,500 women screened over ten years, 1 death from breast cancer had been prevented. Unfortunately, 6 to 10 women were overdiagnosed and overtreated, all no doubt wrongly believing that their lives had been saved by aggressive therapies. The authors concluded, "Overdiagnosis and unnecessary treatment of nonfatal cancer creates a substantial ethical and clinical dilemma and may cast doubt on whether mammography screening programs should exist."

In 2015, researchers from the University of Washington School of Medicine found that for every 10,000 women screened over ten years, as few as 30 and as many as 194 would be overtreated for a cancer that never would have killed them. The authors worried that most women signing up for yearly or biennial mammograms wouldn't know this, and they concluded that "physicians have been more likely to discuss benefits than harms of screening and frequently do not ask about women's preferences."

In 2015, Archie Bleyer updated his estimates of overdiagnosis. Three years had passed since his *New England Journal of Medicine* paper was published. Now he estimated that every year, 78,000 women in the United States were being overdiagnosed and overtreated for a breast cancer that never would have killed them. He lamented that "most patient-education ads do not even mention overdiagnosis, and most women are unaware of its possibilities." He also didn't let doctors off the hook, arguing that by omission, women were being misinformed. "Radiologists, pathologists, and other professionals involved with screening mammography should recognize that the potential harm of overdiagnosis is downplayed or not discussed with the patient and family, despite agreement that the objective is informed choice."

No recent study has ever shown that mammograms have prolonged the lives of American women. This doesn't mean that mammograms don't save some lives. It only means that the number of lives saved by mammograms, given the advances in breast cancer treatments, hasn't been enough to have had an impact on overall mortality.

ALTHOUGH overdiagnosis and the subsequent harms of unnecessary invasive treatments are the greatest problems with early screening, they're not the only ones. Labeled a "false-positive screen," this other problem refers to women who are asked to come back to the doctor due to suspicious findings on the mammogram only to find out later (sometimes much later) that they didn't really have cancer.

In 1996, researchers from Stockholm, Sweden, looked back at a nationwide screening program that had begun in 1981. Thousands of women had been screened. The first round of screening was completed in 1983. The rate of false positives was astounding. The researchers found that 352 women had been diagnosed with a suspicious mammogram. As a consequence, these women made 1,112 repeat visits to the physician, suffered 397 initial biopsies or repeat biopsies, and underwent 187 confirmatory mammograms. After six months, 64 percent of these women were declared cancer-free. For others, it took years before the false alarm was confirmed. Not surprisingly, all these women suffered the anxiety, depression, family disruption, and occasional lost time from work that comes with coping with the notion that they might be suffering from a fatal cancer.

In 1998, researchers from the University of Washington School of Medicine found that among 10,000 women screened with at least one mammogram, 1 in 4 had a false-positive test. After ten mammograms, half had had a false-positive test. These false-positive tests led to 870 outpatient visits, 539 repeat mammograms, 188 biopsies, and 1 hospitalization. The researchers estimated that among 32 million women screened for breast cancer, about 16 million will have had at least one false-positive mammogram. They concluded that "Over 10 years, one-third of women screened had abnormal test results requiring additional evaluation, even though no breast cancer was present . . . Physicians should educate women about the risk of a false positive result of a screening test for breast cancer."

In 2012, researchers from Oslo, Norway, reviewed European breast screening programs involving more than 390,000 women. They found that among women between ages fifty and sixty-nine who were screened for ten years, as many as 82,000 had received a false-positive diagnosis and 23,000 had had an unnecessary biopsy.

FOR those trying to decide whether to undergo screening mammograms, here are the odds.

For every 1,000 women screened every two years, most will have a false-positive mammogram during the next ten years, 146 will have an unnecessary biopsy, 7 will see a fatal case of breast cancer prevented, and 19 will be diagnosed with a cancer that never would have killed them. Among the 19 per 1,000 who are overdiagnosed, 99 percent will have surgery, 70 percent radiation therapy, 70 percent hormone therapy, and 25 percent chemotherapy without benefit.

Because the use of screening mammograms to prevent breast cancer has both risks and benefits, the recommendations from various advisory groups for when or whether to get them are all over the map:

- The U.S. Preventive Services Task Force and the American College of Physicians recommend mammography every two years for women ages fifty to seventy-four.

- The American College of Radiology and the American College of Obstetricians and Gynecologists recommend yearly mammograms for women starting at age forty without an upper age limit.

- The American Cancer Society recommends yearly mammograms for women ages forty-five to fifty-four and biennial mammograms for women age fifty-five and older.

- The Canadian Task Force on Preventive Health Care and the European Cancer Observatory recommend mammograms every two to three years for women between ages fifty and sixty-nine.
- The Cochrane Collaborative and the Nordic Cochrane Collaborative don't recommend mammograms, arguing that the risks outweigh the benefits. Similarly, independent expert groups in Switzerland and France have recommended that breast cancer screening programs should be stopped or substantially reduced.

WOMEN trying to decide whether to undergo annual or biennial mammograms could make one of two choices, both of which are defensible:

> Choice 1: "Among the one thousand women screened, I might be one of the seven who will have had their lives saved by screening. Although I realize that I might be one of the nineteen who will undergo unnecessary surgery, radiation, hormone therapy, and chemotherapy, I'm willing to take that chance."
>
> Choice 2: "I know that women who receive screening mammograms do not live longer than those who don't. I don't want to be one of the many women who suffer either false scares or unnecessary treatments in the hope that I might be the one who benefits."

At the very least, women need to understand the risks and benefits of mammography. Most don't. A recent study found that 69 percent of advertisements for cancer prevention screening programs did not quantify benefits and harms or, worse, presented them in a misleading manner.

SO, what is the answer to the question posed at the beginning of this chapter? How can we tell who will benefit from screening mammograms and who won't? The current answer is that we can't. Until we develop better immunological, biological, or genetic markers that distinguish breast cancers that will kill you from those that won't, women are left with a decision that is based on incomplete information. Probably the biggest offender is ductal carcinoma in situ (DCIS), which accounts for about 20 percent of all breast cancers and might or might not progress to more invasive disease. For this reason, treatment options vary widely. At the moment, there is just no telling who is at risk and who isn't.

There remains no perfect choice, no risk-free strategy.

ONE thing, however, is clear. Once someone has made the choice for surgery, radiation, or chemotherapy, they hold on to that choice with the strength of a religious belief, which is perfectly understandable.

For example, while at dinner with an older couple, I said that I was writing a book that included, among other things, overdiagnosis and overtreatment of prostate, thyroid, and breast cancers—specifically, that many of those diagnosed with these cancers would more likely die with them than from them. Unknown to me at the time, the man had recently had a total prostatectomy and had suffered the inevitable consequence of incontinence, necessitating a pad. He was insistent, however, that his was one of the lives that had been saved. I didn't try to dissuade him. (Had I known that he had been diagnosed with prostate cancer, I never would have brought up the subject.)

For the next few weeks, the man continued to send me emails with commentaries that he believed supported his point of view, that prostate surgery had saved his life. His response isn't surprising. He had made the Big Bet. Once someone has undergone a thyroidectomy or a mastectomy or a prostatectomy, or has suffered through chemotherapy, radiation therapy, or hormone therapy, the only way to reduce the dissonance that they may have made the wrong choice is to believe that what they did was absolutely necessary. Otherwise, all their suffering will have been in vain.

This Big Bet phenomenon was tested years ago at the racetrack. Robert Knox and James Inkster, researchers at the University of British Columbia, went to Exhibition Park in Vancouver and asked bettors to rate their horse's chance of winning thirty seconds before and a few seconds after making a bet on the horse to win. They titled their study "Postdecision Dissonance at Post Time." Knox and Inkster found that bettors consistently rated their horse higher after making the bet than before making it, even though the horse, the jockey, the odds, and the previous track record hadn't changed during the intervening minute. The only thing that had changed was that the bettors had locked in their choice.

PART V
Surgery

14

Heart Stents Don't Prolong Lives

It begins with a crushing pain, like an elephant sitting on your chest. The pain travels to your arm and shoulder, back and jaw. You sweat. You struggle to breathe. You feel nauseated and dizzy. The problem: your heart isn't getting enough oxygen. The medical term is *angina,* from the Greek for "strangling."

THE heart pumps oxygen-containing blood to every organ of the body, including itself. The two main arteries that provide blood to the heart (called coronary arteries) originate from the base of the aorta as it leaves the left ventricle. If these arteries or their branches are blocked, they can't effectively supply oxygen to the heart. If patients have chest pain, they are said to have angina. If the heart muscle is damaged, they are said to have had a heart attack. If the damage is overwhelming, they die.

Heart disease is the leading cause of death worldwide, including in low- and middle-income countries. More people suffer from heart disease than from tuberculosis, AIDS, and malaria combined. By 2030, the worldwide death rate from heart disease will reach 23 million.

In the United States, 17 million people suffer from heart disease, 10 million have angina, and 8 million have heart attacks every year. Indeed, one out of every four deaths in the United States, about 600,000 people, is caused by heart disease—the single most common cause of death in American men and women.

Because of the enormity of this problem, many researchers and clinicians have devoted their professional lives to solving it. Some of the lessons have come at a high price. Some remain unlearned.

The first description of angina appeared in 1772. One hundred years passed before anyone had any idea what caused it. Ludvig Hektoen, an American pathologist, was the first to understand what was happening. Hektoen found that people who died from heart attacks had clogged coronary arteries at autopsy. At the time, no specific therapies for unclogging arteries existed. Rather, those suffering from angina were treated with bloodletting, opium, and bed rest, which weren't particularly helpful. Bloodletting deprived patients of oxygen-carrying red blood cells, and opium launched a generation of opium addicts.

Bed rest, probably the most dangerous of the three, was the only therapy that survived into the twentieth century. From the late 1800s until the early 1960s, patients admitted to hospitals following heart attacks were placed in beds far from nurses' stations so that they wouldn't be disturbed. Removed from constant supervision, many of these patients were found dead in their beds, most from a fatal arrhythmia (a disturbance of electrical impulses in the heart). The mortality rate in people admitted to the hospital following a heart attack was 30 percent. This grim statistic gave birth to coronary care units, where patients' heart rhythms could be monitored constantly with an EKG machine. As a result, the mortality rate dropped by half.

IN the 1960s, two events ushered in the modern era of heart care. Now doctors realized that they could not only treat heart disease but also prevent it.

The National Heart Institute, which was renamed the National Heart, Lung, and Blood Institute, precipitated the first major discovery. In 1948, the institute commissioned experts in the fields of epidemiology, biostatistics, and cardiology to study the lifestyles and habits of five thousand people living in Framingham, Massachusetts. Experts were charged with answering one question: Who was at risk of heart disease? Two decades later, researchers found that those at greatest risk had high blood pressure (hypertension). Clinicians now realized that drugs such as spironolactone (introduced in 1959), propranolol (introduced in 1964), and captopril (introduced in 1980) not only treated hypertension, but also lessened the incidence of heart disease.

The Framingham Heart Study also found something else. Those at risk for severe heart disease had high cholesterol levels. Researchers later learned that not all cholesterols were the same. There was good cholesterol (called high-density lipoprotein cholesterol, or HDL) and bad cholesterol (called very low-density lipoprotein cholesterol, or LDL). At the time that these different cholesterols were being defined, Ancel Keys introduced the concept of the "heart healthy diet." Keys, who chaired the International Society of Cardiology for the World Health Organization, had written several bestselling books on diet and disease. He argued that saturated fats (such as animal fats) were bad and that unsaturated fats (such as those contained in partially hydrogenated vegetable oils) were good. Researchers later found that partially hydrogenated vegetable oils, such as those contained in margarine, were loaded with trans fats, a substance that not only increased the level of bad cholesterol but also decreased the level of good cholesterol. By telling Americans that they were better off eating margarine than butter, Keys had inadvertently driven them into the waiting arms of one of the most dangerous food substances ever ingested. Researchers at Harvard Medical School later estimated that trans fats had caused about 250,000 serious or fatal heart attacks every year.

The finding that high cholesterol levels increased the risk of heart disease gave birth to another class of drugs, statins, the first of which, lovastatin, was introduced in 1987. Like the antihypertension drugs, statins dramatically lowered the incidence of heart disease.

The Framingham Heart Study had taught doctors that lowering blood pressure and lowering bad cholesterol could prevent heart disease. But it was another event that occurred in the 1960s that had the greatest impact. In 1964, the Surgeon General of the United States concluded that cigarette smoking, in addition to causing lung cancer and bronchitis, also caused heart disease. Cigarette smoking can cause high blood pressure, damage the lining of blood vessels, and increase the tendency of blood to clot, all of which can lead to clogged arteries. By 1973, when consumption of cigarettes started to fall in earnest, the incidence of heart disease fell with it. During the past four decades, choosing not to smoke cigarettes has, more than any other medical intervention, been responsible for the 60 percent reduction in the incidence of heart disease.

At this point, researchers knew that coronary arteries could be clogged with cholesterol-containing fatty deposits surrounded by platelets, small cells in the bloodstream that cause clotting. This realization gave rise to another medical treatment: aspirin, an antiplatelet drug that had been around since 1899. Some cardiologists, however, were tired of dancing around the problem. They wanted to open up clogged arteries directly, surgically. On its surface, this idea made a lot of sense. Most people with heart attacks had a greater than 70 percent blockage of only one coronary artery. And the heart muscle that was damaged invariably received its blood supply from that artery. The question at this point was how could doctors get into the artery and what could they do once they got there?

The first person to thread a small catheter into a coronary artery was Werner Forssmann. He did it in 1929. Most impressively, he did it on himself. Forssmann had opened a door. Now cardiologists could theoretically inject a dye into coronary arteries to determine the amount of blockage. Beginning in 1958, they have done exactly that.

The next step was to figure out how to relieve the blockage. The first approach, which was invented in 1976, was to inject a clot-busting enzyme called streptokinase into the blocked artery, which worked. A trial of more than ten thousand people found that streptokinase prolonged the lives of patients who had suffered a heart attack. In 1977, one year after streptokinase was introduced, cardiologists invented balloon dilatation, which involved taking a catheter with a tiny, uninflated balloon at the end of it, threading it into the blocked artery, and inflating the balloon, which opened up the artery.

Nine years later, in 1986, scientists developed bare metal stents. Now cardiologists could place a small meshed metal tube in the blocked coronary artery and leave it there. By 1999, stenting had largely replaced balloon dilatation.

In 2002, another intervention, called drug-eluting stents, was born. This time, the stents weren't made of bare metal only. Rather, they were soaked in drugs that either had an anti-inflammatory effect (sirolimus) or decreased unwanted cell growth around the stent (paclitaxel).

The discovery of risk factors for coronary artery disease, such as high blood pressure, high cholesterol, and cigarette smoking, caused people to suffer less angina, experience fewer heart attacks, and live longer. The same was true for the antihypertension drugs, cholesterol-lowering agents, and streptokinase. The results from these interventions were clear, consistent, reproducible, and compelling. But what about balloon dilatation and stents? Were the results from these surgical procedures also clear? Because threading a catheter into a coronary artery could cause a heart attack or death, both recognized complications of the procedure, they had better be, especially given that tens of thousands of people in the United States and hundreds of thousands in the world were undergoing these procedures every year.

Early studies showed promise. In 1992, researchers from the Veterans Affairs Medical Centers in Massachusetts and Connecticut randomly assigned two hundred patients to receive either medical treatment (such as antihypertension medicines or cholesterol-lowering agents) or balloon dilatation. All the patients suffered from angina, and all had one coronary artery with at least a 70 percent blockage. Six months later, 64 percent of the patients in the balloon group and 46 percent in the medical treatment–only group were free of angina. Patients in the balloon group were also able to increase their time on a treadmill without experiencing chest pain. Two patients in the balloon group, however, had to undergo emergency coronary bypass surgery following a complication from the procedure. The authors of the study concluded, "For patients with single-vessel coronary artery disease, [balloon dilatation] offers earlier and more complete relief of angina than medical therapy and is associated with better performance on the exercise test. However, [balloon dilatation] initially costs more than medical treatment and is associated with a higher frequency of complications."

This study gave birth to the notion that surgical treatment of clogged coronary arteries lessened the frequency of angina, a notion that has been hard to put to rest despite the parade of contradictory studies that followed.

In 2005, researchers from the Netherlands randomly assigned 1,200 patients with stable heart disease to receive either medical therapy or balloon dilatation. Investigators found no differences in the frequency of angina, the rate of hospitalization, or the likelihood of death from heart disease.

Also in 2005, researchers in Greece published a review of eleven studies with a total of 2,900 patients who received either balloon dilatation or medical treatment and found that surgery "does not offer any benefit in terms of death, myocardial infarction [heart attack], or the need for subsequent [surgery] compared with conservative medical treatment."

From this point forward, clinical trials changed in two important ways. Balloon dilatation was replaced by stents, and patients were randomly assigned to receive medical therapy plus stents or medical therapy alone. The results didn't change. Surgery, whether it was balloon dilatation, bare metal stents, or drug-eluting stents, was an unnecessary, expensive, and potentially dangerous procedure.

Heart Stents Don't Prolong Lives

In 2006, researchers from New York University School of Medicine, in collaboration with researchers from around the world, randomly assigned 2,200 patients to receive bare metal stents plus medical therapy or medical therapy alone. They found that stents "did not reduce the occurrence of death, myocardial infarctions, or heart failure."

In 2007, researchers from fifty U.S. and Canadian medical centers randomly assigned 2,300 patients to receive stents plus medical therapy or medical therapy alone, finding that stents "did not reduce the risk of death, myocardial infarctions, or other major cardiovascular events." At the time of this study, more than one million stents had been placed in the United States. A follow-up of this study showed that while those in the stent group were more likely to be free of angina three months after surgery, three years later the benefit disappeared.

In 2009, researchers from Tufts Medical Center, in Boston, reviewed twenty years of studies that initially evaluated balloon dilatation and later bare metal and drug-eluting stents to see if surgery was better than medical therapy alone. It wasn't.

A casual observer at this point would have reasonably assumed that the twenty-year review would have put an end to stenting. It didn't.

In 2014, researchers from the State University of New York Stony Brook School of Medicine reviewed yet another five studies, which involved 5,300 more patients. They found that stents plus medical therapy were "not associated with a reduction in death, non-fatal myocardial infarction . . . or angina compared with medical therapy alone."

Similar to the 2009 review, the 2014 review also didn't eliminate the medical profession's desire to put stents into patients with angina. There would be one more study—a study that should have ended all studies. For years, cardiologists had argued that while stents might not prolong lives or lessen the risk of heart attack, they did seem to lessen the frequency of angina, even if only for a short while and even if only in a few studies. The problem with angina, however, is that it's subjective. People perceive pain differently. Those who opposed stenting argued that following a procedure that is supposed to reduce pain, patients would be more likely to *believe* that they were experiencing less pain—even if they weren't. In other words, the placebo effect. But how could cardiologists perform a placebo-controlled trial of stenting? As it turned out, Rasha Al-Lamee and colleagues at Imperial College London found a way. In 2017, they published the results of their groundbreaking and controversial study.

To control for the placebo effect, the researchers in the United Kingdom randomly assigned patients who had single-vessel coronary artery disease with at least 70 percent blockage to receive a stent or to undergo a procedure where a catheter was threaded up into a coronary artery, but no stent was placed. None of the patients knew whether they had actually received a stent. Patients were then tested on a treadmill to see how long they could exercise before experiencing pain. Researchers found that both groups had improved exercise tolerance, even those with the fake stent procedure.

HEART stents had seemed to make a lot of sense for patients who had previously suffered angina or heart attack. All these patients had single-vessel disease with blockages greater than 70 percent, some as great as 99 percent. After the stent was placed, the blockage was reduced. For patients, the concept of stenting was easy to understand. For doctors, it was easy to explain. So why didn't it work?

The most likely explanation is that a blockage in one large coronary artery was merely indicative of a much bigger problem. Large arteries supply smaller arteries. As it turned out, a blockage in a large artery invariably accompanied blockages in smaller arteries downstream. This explained why relieving the blockage in one large artery didn't solve the problem.

In the end, patients have to ask themselves this question: Should I undergo a procedure that has been shown to be no better than medical therapy alone (after studies involving tens of thousands of patients), is expensive (about $67,000 per procedure), and is potentially dangerous (occasionally causing heart attack and the need for emergency surgery)? Harry Krumholz, a Yale cardiologist, in response to the placebo-controlled study by Al-Lamee and his team, said, "This should make us step back and ask questions about what we are accomplishing with this procedure."

Nonetheless, despite overwhelming evidence to the contrary, every year tens of thousands of Americans with angina receive a stent.

15

Surgery for Knee Arthritis Is Unnecessary

We outlive our knees. Every year, twenty million Americans experience severe knee pain; more than half are younger than sixty-five. About $28 billion a year is spent taking care of this problem. By 2030, seventy million Americans will suffer from knee arthritis.

People with knee arthritis have difficulty running, jogging, hiking, and climbing. In the morning, it takes about thirty minutes for the pain and stiffness to subside. Sometimes, without warning, the knee buckles. In occupations that require bending or lifting, people with knee arthritis can no longer work.

Desperate to do something, anything, to relieve pain and improve function, people take acetaminophen (Tylenol), ibuprofen (Advil, Motrin), naproxen (Aleve), or celecoxib (Celebrex), or they try acupuncture, or they go to health food stores and buy dietary supplements that "support joint health." Surgeons inject steroids into the joint or perform arthroscopic surgery in which, working through a small tube with a fiber-optic light at the end, they wash out the joint (lavage) or trim the cartilage and meniscus (debridement).

Or they perform the definitive procedure: total knee replacement. In the final analysis, however, only one strategy provides the safest, best way to treat this disorder. We'll get to that.

First, we need to understand why knee arthritis is such a universal problem.

Arthritis is the medical term for inflammation of a joint, where one bone meets another. We have joints in our fingers, wrists, elbows, shoulders, hips, knees, and feet. Because the hips, knees, and feet support our weight, they are most vulnerable to the trauma of vigorous activities. No joint bears a larger burden than the knee.

In the knee joint, the thighbone (femur) meets the larger bone in the lower leg (tibia). To provide a cushion, the top of the tibia is coated with a smooth, fibrous layer known as the articular cartilage. This cartilage serves as a shock absorber to prevent the femur from directly rubbing against the tibia. The other shock absorber between the tibia and femur is the meniscus, which further prevents bone from contacting bone. Fluid produced within the knee joint allows for smooth, even movements without pain. Unfortunately, as we get older or heavier or engage in intense physical activities such as football, rugby, soccer, or gymnastics, the cartilage at the top of the tibia thins and weakens. And the meniscus degenerates and frays. Now the femur meets the tibia, which causes the release of tiny fragments of bone, cartilage, and meniscus into the joint. In an attempt to repair the damage, the body sends white blood cells into the area, causing inflammation and pain. The joint space narrows. Ligaments and muscles in and around the joint weaken. The femur and tibia are no longer correctly aligned. All these events set up a vicious circle of more erosion, more inflammation, more pain, more dysfunction, and more immobility.

THE first recorded attempts to relieve the pain of arthritis date back to second-century BC China: acupuncture. The ancient Chinese believed that diseases were treated by allowing the body's vital energy (*qi*) to flow, which involved placing thin needles of between one-half and four inches long under the skin along a series of twelve meridians that arc from head to toe. (Twelve meridians were chosen because China has twelve great rivers.) Because dissection of the human body was illegal in China—and punishable by death—acupuncturists knew nothing about the nervous system. Nonetheless, acupuncture for the treatment of pain has survived well into the modern era, which means it can now be formally studied.

In 2006, researchers from Germany divided one thousand patients into three groups. One group received pain medicines; a second group received traditional acupuncture; and a third group received sham (fake) acupuncture. In the traditional acupuncture group, needles were placed deeply along the meridians prescribed by the ancient Chinese. In the sham acupuncture group, needles were placed randomly and superficially. Ten traditional acupunctures or ten sham acupunctures were performed over six weeks. Those who received traditional acupuncture or sham acupuncture were also allowed to take pain medicines if they wanted.

The researchers found that traditional acupuncture decreased pain, improved function, and lessened the desire for pain medicines; so did sham acupuncture. No difference. The investigators argued that the reason sham acupuncture relieved pain was the placebo effect: it worked because people believed it was supposed to work. What wasn't clear in the German study, however, was whether the placebo effect was psychological or physiological. Because pain is subjective, people may have said they had less pain because they wanted it to be less (i.e., psychological). Unlike blood pressure and temperature, researchers don't have an objective tool to measure pain. But the placebo effect might also be physiological. For example, some studies of acupuncture have shown that Naloxone, a drug that blocks pain-relieving chemicals produced in the brain called endorphins, can block the beneficial effects of acupuncture. In other words, people can learn through acupuncture or sham acupuncture to release their own pain-relieving chemicals. In this instance, the placebo effect was actually physiological.

THE problem with acupuncture, however, is that it doesn't reduce inflammation, or remove tiny fragments of bone, or trim a torn meniscus, or restore frayed or damaged cartilage. So, acupuncture is never going to be a long-term solution to the problem of knee arthritis.

Next up were drugs to treat one feature of arthritis: inflammation. Bayer Aspirin, created in 1897, was the first. Other anti-inflammatory drugs followed. Acetaminophen (Tylenol) was first marketed in the United States in 1955; ibuprofen (Motrin, Advil), in 1974; and COX-2 inhibitors (Vioxx, Bextra, Celebrex), in 1999. Studies have shown that ibuprofen or COX-2 inhibitors work better than acetaminophen or aspirin at relieving the inflammation and pain of arthritis. But like acupuncture, neither of these anti-inflammatory drugs restore eroded cartilage, and all have side effects. Aspirin and ibuprofen can cause inflammation of the lining of the stomach (gastritis) and increase the risk of bleeding. Acetaminophen can cause liver damage (hepatitis). And COX-2 inhibitors are a rare cause of heart attack. But the biggest problem with anti-inflammatory drugs for knee arthritis is that because inflammation recurs, there is no end to taking them.

IN 1950, orthopedic surgeons began injecting steroids into the knee joint. Although steroids also don't address the problem of cartilage erosion, they directly address the problem of local inflammation. Since the inception of this technique, six studies have examined whether steroids effectively treat knee arthritis by comparing people who received steroid injections with those who received saltwater (placebo) injections. The good news was that steroids but not the placebo clearly lessened pain and improved function. The bad news was that the beneficial effect lasted for only a few weeks; three to four weeks later, the benefits of steroid therapy were gone. Because steroid injections don't address the cause of the inflammation, their benefits are transient. As a consequence, steroids, like the other anti-inflammatory drugs, are also not a long-term solution to the problem of knee arthritis. Worse, beginning in the late 1950s, steroid injections were shown to cause cartilage thinning and destruction and to inhibit the cartilage's ability to repair itself. It is, frankly, remarkable that this therapy still survives.

IN the 1970s, with the development of fiber-optic technology, arthroscopic knee surgery was born. Now, instead of putting patients under general anesthesia and making a large incision to open up the entire knee joint, orthopedic surgeons could make several tiny incisions, each less than half an inch, and insert a fiber-optic arthroscope, an outpatient procedure that could be done under local anesthesia. Once the arthroscope was inserted, surgeons could lavage (wash out) the joint with about ten quarts of saltwater. This enabled them to remove large and small pieces of cartilage, bony fragments, calcium crystals, and inflammatory cells. Surgeons could also debride the joint by smoothing out the frayed articular cartilage and meniscus, thus interrupting a process that produced more fragments, more inflammation, and more pain.

By the turn of the twenty-first century, 650,000 arthroscopic surgeries of the knee were performed in the United States every year and 2 million worldwide. Using a procedure that was safe and easily performed, surgeons could now finally do something that addressed the central problem of knee arthritis: the fraying of the articular cartilage and meniscus. The vicious circle could now be broken. By 2002, fourteen studies had shown that arthroscopic surgery offered substantial pain relief. None, however, compared people who had had arthroscopic surgery with those who hadn't.

Arthroscopic Surgery Isn't Better Than Physical Therapy for Knee Arthritis

In 2002, researchers from the Houston Veterans Affairs Medical Center and Baylor College of Medicine performed what was at the time a groundbreaking and, to some, unethical study. The investigators assigned 180 patients to one of three groups. Each group was brought into the operating room, where three incisions were made in the knee. For the first group (called the lavage group), a fiberoptic arthroscope was inserted and the knee was washed with ten quarts of saltwater. For the second group (the debridement group), rough articular cartilage was shaved, loose debris was removed, torn or degenerated meniscal fragments were trimmed, and the remaining meniscus was smoothed to a stable rim. It was the third group (the placebo group) that raised eyebrows. After three incisions were made in the knee, patients in the placebo group heard the surgeon ask for instruments, heard splashing sounds simulating lavage, stayed in the operating room for as long as patients in the other two groups, and, as with the lavage and debridement groups, spent a night in the hospital. The difference for the placebo group was that nothing had been done to them. No insertion of an arthroscope, no lavage, no debridement. All the placebo patients thought that they had just undergone arthroscopic knee surgery when none of them had.

To control for possible differences in techniques among surgeons, only one surgeon performed all these procedures. This particular surgeon had been in practice for ten years, was board-certified and fellowship-trained in sports medicine, and was the orthopedic surgeon for an NBA team as well as for the 1996 men's and women's Olympic basketball teams. Surgical technique, therefore, wasn't an issue. The only major limitation of this study was that it was performed at a Veterans Affairs Medical Center, which at the time meant that virtually all the participants were men.

During the next two years, the researchers evaluated patients in each of the three groups for knee pain and function. The results were published in the prestigious *New England Journal of Medicine*. There was no difference. In other words, thinking that you had had arthroscopic knee surgery was as good as actually having had it.

How was this possible?

In an accompanying editorial, researchers from Boston University and the University of Iowa offered an explanation. "Although smoothing cartilage and meniscal irregularities may sound appealing," they wrote, "larger forces within and outside the joint environment, such as malalignment, muscle weakness, instability, and obesity, which are not addressed by this type of surgery, may have greater effects on the clinical outcome . . . *Debridement and lavage may simply remove some of the evidence while the destructive forces of [knee arthritis] continue to work*" (emphasis mine). In other words, arthroscopic surgery had treated the symptoms of knee arthritis, not the problem. As a consequence, the problem persisted, and the symptoms recurred.

A series of studies confirmed these results.

In 2007, researchers from Sweden divided one hundred middle-aged patients with knee arthritis into two groups. One group received arthroscopic surgery plus physical therapy; the other, physical therapy alone. The physical therapy program, which consisted of stationary bicycling, stair walking, jogging, jumping, stretching, and knee flexions and extensions, was performed twice a week for eight weeks. Six months later, researchers evaluated knee pain and function. Both groups improved, and both groups were satisfied with the improvement. Again, no difference.

In 2008, researchers from Canada divided eighty-six patients with knee arthritis into two groups. One group received debridement and lavage of the knee joint plus physical therapy; the other, physical therapy alone. Three months later, the arthroscopy group had less knee pain and better knee function than the physical therapy–only group. Two years later, however, the groups were indistinguishable. And unlike the Veterans Affairs Medical Center study, this study included both men and women.

One question that hadn't been answered by these studies, however, was what to do about people with meniscal tears. Surely, these patients would benefit from arthroscopic surgery. As it turns out, they didn't.

In 2013, researchers from Sweden divided one hundred middle-aged men who had degenerative meniscal tears into two groups.

One group received arthroscopic surgery plus exercise therapy; the other, exercise therapy alone. Two and five years later, investigators performed tests of knee pain and function. No difference. Even for people with severe meniscal tears, a commitment to exercise was good enough.

Also in 2013, researchers at Brigham and Women's Hospital, in Boston, in collaboration with researchers at Washington University in St. Louis, the Cleveland Clinic, Vanderbilt University, and the Mayo Clinic, performed a study that should have ended the debate about arthroscopic surgery for meniscal tears once and for all. The investigators divided 350 patients with knee arthritis and degenerative meniscal tears into two groups. One group received surgery plus physical therapy; the other, physical therapy alone. The exercise program focused on balance, range of motion, muscle strength, muscle length, and aerobic conditioning. Evaluations six and twelve months later revealed no differences between the two groups. Surgery to repair degenerative meniscal tears was an unnecessary, costly procedure.

Nonetheless, the studies continued.

In 2018, researchers in the Netherlands completed a two-year study in which they divided 320 patients into two groups. One group had arthroscopic surgery; the other, sixteen exercise sessions for eight weeks. They found no differences in knee function.

In total, fourteen randomized, controlled clinical trials and twelve observational studies, involving 1.8 million people, found that arthroscopic surgery for knee arthritis, with or without repair of a torn meniscus, was no better than physical therapy alone. Arthroscopic surgery, therefore, is no longer recommended for the treatment of knee arthritis. Yet, it remains one of America's most common outpatient surgical procedures.

NONE of the surgeries or medicines just described solved the central problem of knee arthritis, the constant fraying and loss of articular cartilage. Because patients have been generally frustrated by the persistence of symptoms despite medical interventions, they've become easy prey for hucksters.

This brings us to the dietary supplement industry.

For decades, supplement makers have claimed that because glucosamine and chondroitin sulfate are components of cartilage, the combination of the two can cure joint pain. In the United States, 6.5 million adults consume these two supplements every year. Apart from patient and celebrity endorsements, is there any evidence that they work?

In 2006, researchers from the University of Utah School of Medicine, in collaboration with researchers from across the country, divided 1,600 patients with knee arthritis into five groups. One group received 1,500 milligrams of glucosamine daily for twenty-four weeks; the second group received 1,200 milligrams of chondroitin sulfate; the third group received the two supplements in combination; the fourth group received 200 milligrams of Celebrex, an anti-inflammatory drug; and the fifth group received placebo pills. At the end of the study, glucosamine and chondroitin sulfate, given either separately or together, were no better at reducing pain or improving function than placebo pills. The only patients whose improvement was significantly better than that of placebo recipients were those who had received the anti-inflammatory drug.

Not only did the ingestion of glucosamine and chondroitin sulfate not work to treat knee arthritis, but it never made sense that it would. While it is true that these substances are components of cartilage, supplement makers were asking consumers to believe a series of impossible events. Once swallowed, glucosamine and chondroitin sulfate would have to survive the harsh acid environment of the stomach, travel directly to the affected joint, and significantly enhance the formation of new cartilage. But the problem with knee arthritis isn't that the body can't make new cartilage; we make an abundance of new cartilage every day. The problem is that poor alignment of the femur and tibia caused by weakened ligaments within the joint, weakened muscles around the joint, and the trauma of walking or standing or being obese are constantly fraying new cartilage. The surprise of the glucosamine–plus–chondroitin sulfate experiment wasn't that it hadn't worked; it's that it would have been remarkable if it had.

ONE surgical procedure, however, cures knee arthritis—a procedure that eliminates the problem of joint misalignment, fraying cartilage, the accumulation of bony fragments and crystal deposits, and the inevitable inflammation, pain, and immobility that follow; a procedure that, like knee arthroscopy, was first introduced in the 1970s; and a procedure that is now performed more than 670,000 times every year in the United States at a cost of about $36 billion: total knee replacement. It took forty years for researchers to finally study total knee replacement plus physical therapy to see if they worked better than physical therapy alone. When these studies were first performed, they were called parachute studies, the implication being that you don't need to do a placebo-controlled study of parachutes when someone jumps out of an airplane any more than you need to do a placebo-controlled study of antibiotics for meningitis or insulin for diabetes or chemotherapy for treatable cancers. You know that these therapies are going to work, so why subject placebo recipients to something from which they clearly won't benefit? As it turns out, two large studies of total knee replacement proved that these weren't parachute studies after all.

In 2015, researchers from Denmark divided one hundred patients with knee arthritis into two groups. One group received a total knee replacement followed by twelve weeks of nonsurgical therapy; the other group, nonsurgical therapy alone. Nonsurgical therapy, which was delivered by physiotherapists and dieticians, consisted of exercise, education, dietary advice, foot insoles, and pain medication. One year later, the researchers found that 85 percent of those who had received a total knee replacement had less pain and better function. What was interesting, however, was that 68 percent of the nonsurgical group also experienced significant improvement—so much so, in fact, that when given the opportunity to undergo a total knee replacement during the study, only 26 percent of those in the nonsurgical group chose to do it. The rest were happy with the outcome without surgery. Also, the surgery group had a much higher rate of serious side effects, including infection, deep vein blood clots, fractures of the bone above the knee replacement, and stiffness of the new joint so severe that it required manipulation of the knee in a hospital under general anesthesia.

In 2018, these same researchers completed another trial, this time with two hundred patients. The results were the same.

FOR the past forty years, orthopedic surgeons and internists have offered a variety of treatments for arthritis of the knee. Ironically, the one therapy that is both the least expensive and the safest appears to be the best: physical therapy.

As a general rule, strength and conditioning programs are eight weeks long, include three sessions per week, and last thirty minutes per session. These programs feature flexion and extension against resistance and low-impact aerobic exercises designed to decrease pain, increase range of motion, and increase strength. Patients also receive education on posture and gait and how to improve physical fitness and reduce weight. Most important, these programs address the main reason that cartilage continues to deteriorate: malalignment of the joint. By strengthening the ligaments within the joint and the muscles around the joint, the persistent erosion of cartilage lessens, as do inflammation, pain, and restricted function.

In the hands of the right physiotherapist, these programs can even be fun. Tai chi, which is characterized by soft, slow movements that emphasize force rather than brute strength, can be done in groups; as can yoga, which emphasizes emotional and spiritual well-being. Even simple daily exercises such as walking, gardening, swimming, and dancing can be structured to improve the symptoms of arthritis.

Best of all, not only do these programs avoid the moderate side effects of pain medicines; they avoid the cost, debilitation, and occasionally severe and even fatal consequences of major surgery.

16

Don't Remove Mercury Dental Fillings

The dose makes the poison.

—PARACELSUS, SIXTEENTH-CENTURY SWISS PHYSICIAN

If you have teeth, you've probably had cavities. It's easy to see how this can happen. Tooth decay begins with bacteria in plaque, a thick substance that sticks to teeth and gums. These bacteria interact with starches and sugars in food to erode the outer surface of the tooth, causing holes (cavities). About 40 percent of children will have cavities in their primary teeth, and 90 percent of adults will have them in their permanent teeth. If untreated, cavities can cause pain, inflammation, and tooth loss.

To treat cavities, dentists drill out the decayed material and fill the hole with either a composite resin or an amalgam. Composite resins, which are the color of teeth, are made from the sap of trees, notably fir and pine. Amalgams are made by mixing powdered tin, copper, and silver with liquid mercury (otherwise known as elemental mercury or quicksilver).

Although amalgams are often referred to as "silver" fillings because of their color, they are 50 percent mercury. Because they are strong, durable, long-lasting, inexpensive, and can be placed more quickly than resins (an advantage when working with children), dentists have been using mercury-containing amalgams since the nineteenth century. Mercury also slows the growth of bacteria, making it less likely that the tooth will erode further. The Food and Drug Administration estimates that between 1988 and 2008, dentists placed more than one billion mercury amalgams and that they are placing one million more every year.

DESPITE their many advantages, mercury amalgams have always been controversial. On December 3, 1834, fifteen prominent dentists in New York City founded the American Society of Dental Surgeons (ASDS), the nation's first dental association. Their first act was to declare mercury fillings unsafe and force members to sign a pledge promising never to use them. Those who refused were threatened with malpractice lawsuits. This was the beginning of what has been called "the amalgam wars." Membership declined. By 1856, the ASDS disbanded, replaced three years later by the American Dental Association, which has consistently declared mercury amalgams to be safe.

Fears of mercury fillings abated. Then, in the 1970s, they resurfaced courtesy of a dentist from Colorado. Hal A. Huggins graduated from the University of Nebraska School of Dentistry in 1962. In 1985, Huggins and his wife, Sharon, published *It's All in Your Head,* which claimed that mercury fillings were toxic. Huggins had been inspired by an Argentinian dentist who told him that mercury amalgams caused leukemia and Hodgkin's disease. Huggins took it a step further, claiming that amalgams also caused depression, anxiety, irritability, facial twitches, muscle spasms, epilepsy, multiple sclerosis, rapid heartbeat, arthritis, rheumatic diseases, allergies, ulcers, cancer, Crohn's disease, and mononucleosis. (In other words, pretty much everything.) Huggins argued that if people wanted to alleviate these symptoms or cure these diseases, all they had to do was remove their fillings. On May 1, 1996, the Colorado Board of Dentistry revoked Huggins's license. On November 2, 2008, the board issued a cease and desist order when it found that he was still seeing patients.

Nonetheless, Hal Huggins had created a following, and it's easy to understand how this could have happened. Mercury *is* toxic. And you don't have to look very far to prove it. Two events that occurred in the 1950s and '70s showed just how toxic mercury could be.

ON April 21, 1956, a five-year-old girl from Minamata, a town on the west coast of the island of Kyushu in Japan, developed high fever; tremors; sensory disturbances; difficulties walking, speaking, seeing, and hearing; seizures; paralysis; and loss of consciousness. She lapsed into a coma, and finally died. Doctors couldn't figure out what was wrong. Then her younger sister developed the same symptoms. The girls' mother told the doctors that her daughters weren't alone: a neighbor's child was also suffering from this same unexplained illness. A house-to-house search of the neighborhood uncovered eight more patients. On May 1, the director of the local hospital alerted public health officials that the town of Minamata was experiencing an "epidemic of unknown disease of the central nervous system." By the end of May, thirteen more patients from villages near Minamata were discovered. All of them, like the little girl, died. The disease also affected pregnant women; within months, hundreds of babies were diagnosed with cerebral palsy.

This mysterious illness wasn't limited to people. Fish began to float on the surface of Minamata Bay. Then cats started jumping into the sea and dying; local residents called them "cat suicides." Seagulls fell from the sky. Seaweed stopped growing in seabeds.

On November 4, a research group from the local university declared that the disease was caused by heavy metal poisoning. The problem was eventually traced to the Chisso chemical company. In August 1951, five years before the appearance of this devastating illness, the company had begun dumping large quantities of mercury into the bay. For years, local residents had unknowingly consumed fish and shellfish that were heavily contaminated with mercury. Twelve years passed before Chisso finally took responsibility for its actions and stopped dumping mercury. But it was too late. By the time the dust settled, 2,200 people had been poisoned and 900 had died.

The Minamata Bay disaster was the worst environmental mercury poisoning in history. But it wasn't the only one.

In the fall of 1971, 93,000 tons of wheat and barley were shipped from Mexico and the United States to Basra, a city in Iraq. To prevent contamination, the grain had been fumigated with methylmercury, a fungicide. Because it had been fumigated, the grain wasn't meant for consumption, as had been carefully spelled out on the sacks. It was only to be used for seed. Unfortunately, the warnings were written in Spanish and English, not Arabic. Residents didn't know that the grain had been fumigated. Also, the skulls and crossbones on the sacks meant nothing to the Iraqis. Not understanding the warnings, local residents used the grain to make bread—hundreds and hundreds of loaves of bread.

The first case of mercury poisoning occurred on December 21. By December 26, the hospital had alerted the government. By January 1972, the Iraqi Army ordered immediate disposal of the grain and its products; anyone found to be selling it would be put to death. But by then, as had been true with the Minamata Bay disaster, it was too late. More than 6,500 Iraqis were admitted to the hospital, and 460 died. Many Iraqi doctors believed that these estimates were far too low, and that as many as 100,000 people suffered permanent brain damage from the incident.

WHILE the Minamata Bay and Iraqi fumigated grain disasters were avoidable, mercury, which is part of the earth's crust, isn't. Everyone living on the planet is exposed to mercury every day. Nonetheless, most people don't suffer mercury poisoning. Why?

Mercury in the earth's crust is typically combined with sulfur and oxygen. In this form it's called inorganic mercury. For the most part, inorganic mercury is harmless. The problem comes when inorganic mercury is liberated from the earth's crust by either natural processes, such as rock erosion and volcanoes, or by man-made processes, such as burning coal. Every year, between 4,700 and 9,000 tons of mercury are released into the atmosphere. Once released, mercury can be taken up by bacteria in the soil or seas and methylated. Methylation is the process of adding one carbon and three oxygen atoms; the final product is called methylmercury. Unlike inorganic mercury, methylmercury, also known as organic mercury, is far from harmless. (The term *organic* has evolved over the years. Today, when people say "organic," they usually mean free from pesticides, fertilizers, and other artificial agents. Chemists, on the other hand, use the word *organic* simply to mean compounds that contain carbon. Because methylmercury contains carbon, it is called organic mercury.)

Once methylated, mercury enters the water supply and eventually the food chain. For humans, the most common source of mercury is fish. If ingested, methylmercury, unlike inorganic mercury, readily crosses the lining of the intestine and enters the bloodstream. From there, it can enter the brain or kidneys and cause significant damage. Both the Minamata Bay disaster and the Iraqi fumigated grain disaster were examples of methylmercury poisonings.

Because they sit atop the ocean's food chain, predatory fish, such as swordfish, tuna, shark, king mackerel, and tilefish, contain the most methylmercury. Anyone who eats large quantities of these fish might be at risk of subtle symptoms of mercury poisoning. No one is more susceptible than the unborn child. For this reason, pregnant women are asked to consume no more than six ounces of tuna and no more than twelve ounces of other fish or shellfish per week.

Because methylmercury is in the water supply, it's unavoidable. Anything made from water on this planet, including breast milk and infant formula, contains methylmercury. For example, a breastfed child will ingest about 360 micrograms of methylmercury in the first six months of life. (A microgram is one-millionth of a gram.) Although that might sound like a lot, it isn't. That quantity is well within the limits of what is considered safe. If it weren't, everyone on earth would be poisoned by mercury.

ALTHOUGH people are most likely to be harmed by mercury when they eat it or drink it, that's not the only way it can enter the bloodstream. Probably the most well-known example can be found in Lewis Carroll's *Alice's Adventures in Wonderland*. Nitrous mercury, which was used to make felt hats well into the twentieth century, contains nitrogen, not carbon. Because it's inorganic (i.e., it doesn't contain carbon), it can't be absorbed through the intestine. When inhaled, however, nitrous mercury vapor enters the lungs and then the bloodstream. People who worked in the haberdashery industry often suffered from mental illness caused by chronic mercury poisoning. This was the source of Lewis Carroll's Mad Hatter.

Mercury dental fillings, which are made from elemental mercury, are another form of inorganic mercury. Although mercury in these fillings isn't linked to carbon, and therefore can't enter the body through the intestine, it can be vaporized after brushing teeth or eating acidic foods. So, the critical question is: Does mercury vapor from amalgams cause harm?

The World Health Organization has adopted a safety limit for mercury vapor of 25 micrograms per square meter of air per day. Given that symptoms of mercury toxicity have never been observed below exposures of 100 micrograms of mercury per square meter per day, the WHO guideline provides a margin of safety. The quantity of mercury found in the mouths of people who have at least eight mercury fillings is about 1 microgram total per day, at least 100 times less than levels found to be harmful. To put this in perspective, people typically eat or drink about 3.5 micrograms of methylmercury every day. Therefore, the quantity of methylmercury that we consume is more than three times greater than the quantity of mercury vapor we inhale from mercury fillings. And both these sources of mercury are well within ranges shown to be safe. Nonetheless, because people had become concerned that mercury fillings might cause harm, researchers studied it.

In 1988, investigators from Sweden gave one thousand women questionnaires containing an exhaustive list of symptoms, including dizziness, poor vision, hearing defects, headache, general fatigue, sleep disturbances, nervous complaints, sweating, breathlessness, chest pain, cough, irritability, overexertion, inability to concentrate, restlessness, depression, easy crying, abdominal pain, diarrhea, constipation, poor appetite, weight loss, weight gain, sensitivity to cold, difficulty urinating, joint complaints, and back and leg problems. The researchers found no correlation between the number of mercury amalgams and any of these symptoms.

THEN the American media weighed in. On December 23, 1990, *60 Minutes* aired "Is There Poison in Your Mouth?" Morley Safer was the host. He opened the program by scaring the hell out of his viewers. "Is there poison in your mouth?" he asked. "The Dental Association says there isn't. But some of its members say there is and have stopped using it. *It* is a filling. A silver amalgam filling. The dentists' filling of choice for more than a century. More than a hundred million of them were put into American mouths last year. What you probably don't know is that these so-called silver fillings are 50 percent mercury. And mercury is more poisonous than lead or even arsenic. Because it's been around so long and because it was assumed that the mercury was made stable when mixed with other metals, amalgam fillings were never tested for safety . . . But now a growing number of scientists, doctors, and dentists are saying silver amalgams should be banned!"

Hearkening back to the days of the amalgam wars, and ignoring everything known about elemental mercury, a dentist explained to Safer that there was "no safe threshold for mercury exposure." His declaration was followed by a parade of anecdotes.

Safer interviewed a woman who described how she had been "crippled by arthritis, fatigue, and memory loss." Her doctor told her that in six months she would be confined to a wheelchair. But after her fillings were removed, she threw away her cane and had more energy; also, her tremors stopped. "I could do things with my hands again," she enthused, failing to mention that at the same time a rheumatologist had prescribed an anti-inflammatory drug.

A patient with multiple sclerosis said that the morning after the removal of five mercury fillings, she, too, threw away her cane. "It was that quick?" asked Safer. "It was that quick," she replied. "[The dentist] knew it would be of benefit, but no one expected it would be instantaneous or so dramatic. My voice came back. My ability to walk and hold a pencil came back . . . That night I ended up going to dinner in San Francisco and actually [ended up] dancing two dances. And I hadn't been walking since May!" Safer failed to challenge the obvious flaw in her story: During removal of mercury fillings, mercury is vaporized and inhaled into the lungs, where it immediately enters the bloodstream. As a consequence, all patients who have their mercury fillings removed will initially *increase* their blood mercury levels, not decrease them. If mercury poisoning really caused multiple sclerosis (which it doesn't), the woman's symptoms should have worsened immediately after the procedure, not improved.

Safer then interviewed a spokesperson from the American Dental Association (ADA), who explained that the quantity of mercury vapor released from amalgams was "far below anything that would cause a problem." He also issued a warning: "The National Multiple Sclerosis Society has stated that this is a cruel hoax on these people to take the fillings out and hope it's going to cure the MS," he said. "It does not happen in those cases. It simply does not happen."

By interviewing someone from the ADA, Safer had provided the appearance of balance. But this wasn't a balanced program. Safer's point of view was clear. Mercury amalgams were causing harm. "There are thousands of anecdotal stories all over the United States," he explained, showing video of mercury amalgam sufferers who now, after removal of their fillings, finally had relief from "manic depression, chronic fatigue, migraine headaches, multiple sclerosis, and anemia."

If all these people were benefiting from removal of their mercury fillings, why weren't dentists or doctors or scientists or public health officials sounding the alarm? The dentist interviewed in the

60 Minutes program (who was cast as a whistle-blower) explained why. He said that the American public was the victim of a massive conspiracy to hide the truth. "The FDA committees consist of dentists who use it and manufacturers who make it," he said. "There is no basic science input on that committee. Anything the ADA wants it pretty much can get through the FDA. That's what's called effective lobbying." Dentists were making money. Amalgam manufacturers were making money. And no one seemed to care that at the same time, they were sacrificing the health of the public—at least, not until *60 Minutes* told its viewers what was really happening, or so it seemed.

Later, *Consumer Reports* published a letter it had received following the *60 Minutes* program: "My mother, who was diagnosed with Lou Gehrig's disease more than two years ago, had her mercury fillings removed immediately after the show aired. After she spent $10,000 and endured more than 18 hours of dental work [that was] so painful she once fainted in the waiting room, her condition did not improve. The pain was outweighed only by the monumental disappointment she and the whole family experienced as we lived through one false hope." The same year, *Consumer Reports* wrote, "The false diagnosis of mercury-amalgam toxicity has such a harmful potential and shows such poor judgment on the part of the practitioner, that *Consumer Reports* believes that dentists who engage in this practice should have their license revoked."

Stephen Barrett of Quackwatch has called the amalgam controversy one of "the greatest unfounded health scares of modern times."

AFTER the *60 Minutes* program heightened public fears about mercury fillings, millions of dollars were spent by dentists, scientists, and public health advocates to determine whether these concerns were valid. Was it possible that mercury fillings were causing harm? About two dozen studies have now shown that they weren't.

Mercury Fillings Don't Cause Neurological Problems in Adults

In 1991, researchers from the University of Kentucky correlated the number of mercury fillings with tests of cognitive function in 130 Roman Catholic Sisters. They found no relation between the number of fillings and tests of memory, language, visual and spatial ability, concentration, or orientation.

In 2005, researchers from the National Institutes of Health in Bethesda studied the relationship between dental amalgams and neurological symptoms in 1,700 members of the United States Air Force who had served during the Vietnam War. They concluded that "no significant associations were found between dental amalgam exposures and clinical neurological signs of abnormal tremor, coordination, gait, strength, sensation or muscle stretch reflexes."

Mercury Fillings Don't Cause Multiple Sclerosis

In 1998, researchers from the University of Montreal interviewed 270 people who did or did not have multiple sclerosis. The number and duration of mercury fillings were the same in both groups.

In 2004, researchers from New Zealand reported a study of twenty thousand people from the New Zealand Defence Force performed between 1977 and 1998. The investigators found no relationship between the number or duration of mercury fillings and multiple sclerosis, kidney disease, or chronic fatigue syndrome.

Mercury Fillings Don't Cause Alzheimer's Disease

In 1999, researchers from the University of Kentucky compared the number and duration of mercury fillings in sixty-six people with Alzheimer's disease with thirty-three people without the disease. In subjects who died during the study period, autopsies were obtained to determine mercury levels in the brain. The authors found "no significant association of Alzheimer's disease with the number, surface area or history of dental amalgams [and that] dental amalgam restorations, regardless of number . . . do not relate to brain mercury levels."

Mercury Fillings Don't Cause "Amalgam Syndrome"

During the *60 Minutes* segment, Morley Safer interviewed one victim who said that her fillings had caused her to suffer arthritis, fatigue, memory loss, and tremors. This collection of vague symptoms is often referred to as amalgam syndrome. Researchers have conducted three studies to determine if such a syndrome actually exists.

In 1996, investigators from the Department of Dental Material Sciences at Umeå University in Sweden compared ten patients with subtle symptoms believed to be caused by mercury poisoning from fillings with eight people who also had fillings but didn't have these complaints. The authors found no differences between the two groups in either the quantity of mercury vapor released from the teeth or the levels of mercury in the blood or urine.

In 2001 and again in 2008, researchers from Germany compared 107 patients with complaints they believed were associated with their amalgams with 107 patients with a similar number of amalgams who didn't have these complaints. They found no differences in the amount of mercury in their blood, saliva, or urine.

Mercury Fillings Don't Affect the Immune System

In 1991, researchers from the Medical College of Georgia compared twenty-one subjects with amalgams with sixteen without amalgams, finding no differences in levels or distribution of immune cells. They concluded, "The results of this study show no indication that amalgam restorations affect the human immune system."

Mercury Fillings Aren't Harmful to Children

In many ways, the best studies of mercury fillings were those in which composite resins were used as controls. All these studies were performed with children.

In 2008, American and European researchers studied five hundred children in Lisbon, Portugal. The children were divided into two groups. One group received mercury amalgams; the other, composite resins that didn't contain mercury. During the next seven years, the researchers found no differences in memory, attention, visual function, motor function, school performance, or nerve conduction velocities. The authors concluded that "amalgam should remain a viable dental restorative option for children."

In 2008, researchers from Harvard Medical School divided 530 children between ages six and ten into two groups. One group received mercury amalgams; the other, composite resins. After five years, the investigators found no differences in tests of intelligence, achievement, language, memory, learning, visual skills, spatial skills, verbal fluency, fine motor function, problem solving, or attention. They also found no differences in delinquent behavior, aggression, school adjustment, emotional symptoms, or IQ scores.

In 2010, researchers from the University of California, Davis compared blood levels of mercury in three groups of children between ages two and five: those with autism, those with autism spectrum disorder, and those who were developmentally normal. They found that blood levels of mercury were similar among the three groups. Despite concerns by some parents, mercury was not associated with autism.

Mercury Fillings Placed During Pregnancy Don't Affect the Unborn Child

As was found in both the Minamata Bay and Iraqi fumigated grain disasters, the placenta is not a barrier to mercury. In both cases, large quantities of methylmercury ingested by pregnant women damaged their unborn children. For these reasons, several European countries, exercising an abundance of caution, recommend that pregnant women should not receive mercury amalgams during pregnancy. Careful studies, however, don't support this recommendation.

In 2005, researchers from Italy examined the relationship between the number and duration of mercury fillings, the levels of mercury in amniotic fluid, and the outcomes of seventy-two pregnancies. They concluded that "mercury levels detected in amniotic fluid were low and no adverse outcomes were observed through pregnancies and in the newborns."

In 2005, researchers from the University of Washington in Seattle studied 1,100 women with low-birth-weight infants and compared them with 4,500 women whose children were of normal weight at birth to see if there were any differences in the number of amalgams placed during pregnancy. There weren't. The authors concluded, "This study found no evidence that mercury-containing dental fillings placed during pregnancy increased low-birth-weight risk."

In 2007, researchers from the University of North Carolina studied the effect of mercury fillings placed in 7,400 women before or during pregnancy on their children. They found no relationship between umbilical cord mercury levels and development in children up to fifteen months of age, concluding, "Overall dental care, including amalgam fillings, was not associated with birth outcome or language development."

In 2008, researchers from the University of California studied the relationship between the number and timing of dental amalgams and the amount of mercury in maternal and cord blood. The authors concluded that "none of the cord blood mercury concentrations reached the level considered to be hazardous for neurodevelopmental effects in children exposed to mercury [during pregnancy]."

AS a consequence of these and other studies, every major advisory body concerned with the public's health has reaffirmed the safety of mercury amalgams. The World Health Organization, the U.S. Department of Health and Human Services, the European Commission, and Health Canada have stated that "no scientific evidence supports the association between amalgam mercury exposure and adverse events in adults or children." The ADA's Council on Ethics, Bylaws, and Judicial Affairs took this recommendation one step further, stating, "removal of amalgam restorations solely for the alleged purpose of removing toxic substances from the body, when such treatment is performed at the recommendation of the dentist, presents a question of fraud or quackery."

By 2010, studies of mercury fillings had shown that they hadn't harmed adults, children, pregnant women, or unborn children. Toxicological studies had shown that the quantity of mercury released from fillings was well within limits considered safe and that mercury exposure from fillings was insignificant when compared with other sources of environmental mercury. The science was settled. Mercury fillings were safe.

ENTER Dr. Oz. On March 28, 2013, *The Dr. Oz Show* aired a segment titled "Toxic Teeth: Are Mercury Fillings Making You Sick?" (a title that appears to answer its own question).

Oz opened the show holding a mercury thermometer. "Our show today concerns every one of you watching who ever had a cavity filled. Because if your filling is made with silver, it probably also contains the highly toxic element mercury. How toxic? This thermometer contains mercury. If I were to drop it, we would have to evacuate this entire studio immediately." (Broken mercury thermometers require careful cleanup, not building evacuations. Given Oz's warning, it's amazing that those of us who played with those silvery globs of quicksilver in chemistry sets ever survived childhood.)

Oz continued: "So why are dentists still putting fillings containing mercury in your mouth? And if they're already there, could your fillings be poisoning you and making you sick? The American Dental Association insists your fillings are safe. But now there is mounting new evidence showing mercury is released when you eat and drink, and even when you brush your teeth." (New evidence? Studies showing that small quantities of mercury vapor are released in people with amalgams have been available to the public for thirty years.)

Oz then interviewed an "integrative" dentist who offered a series of misleading statements: "This is one of the most hotly contested conversations going on in dentistry today." (Mainstream dentists don't debate this issue, because it's not debatable.) "There are not enough controlled studies." (Apparently more than two dozen controlled studies aren't enough.) "We don't know how much [mercury] makes you sick." (Following the Minamata Bay and Iraqi fumigated grain disasters, we now know exactly how much mercury makes you sick.)

"But what about people who believe they have poisoning from mercury in their amalgams?" asked Oz. "I know there is some research showing people who have had them removed showing improvement."

Channeling Morley Safer, and in direct contradiction to every major public health agency, including one that called the practice "fraud or quackery," Oz then argued that removing mercury amalgams could treat a variety of illnesses. To prove it, he did what Morley Safer did; he relied on anecdote, interviewing a woman in the first row of his studio audience. "When I was younger, I had about fifteen mercury fillings placed in my mouth," she said. "And as I got older, I started to have terrible tremors, hand tremors. I couldn't concentrate, and I thought I was losing my memory. I know it sounds crazy." "I hear these complaints from a lot of people," offered Oz. The guest then said that following removal of all fifteen fillings, "Some of my symptoms went away immediately." (Again, initial removal of mercury fillings will only increase blood mercury levels.)

To prove that mercury fillings were dangerous, Oz performed a science fair project for his audience. He walked over to a glass box containing a set of plastic teeth with mercury fillings. Using rubber gloves, he reached into the sealed box and brushed the teeth. The box was connected to a mercury vapor detection unit, which immediately registered mercury in the air. Oz's colleague stated that any amount of mercury more than zero micrograms was toxic, and that 61 micrograms had been released. But a closer look at the recording device showed that the value wasn't 61; it was 0.061 micrograms per square meter, well within levels shown to be safe. Triumphant, Oz said, "So how can anyone dispute that mercury is coming off amalgams?" (No one has.)

Now that Oz had thoroughly scared his viewing audience, he turned back to the integrative dentist for help. "Should you have your mercury fillings taken out?" he asked. The dentist responded that you "should take them out if you have eight or more fillings, grind your teeth, eat acidic foods regularly, or drink soda regularly." (This would apply to millions of people.) The dentist further offered that people with mercury fillings should avoid acidic foods like coffee, vinegar, grapefruit, and tomatoes. (In other words, everyone.)

Then, as is often true at the end of many of his shows, Mehmet Oz offered a magical dietary supplement. In this case, chlorella, which he claimed would eliminate mercury through the intestines. (And which was, according to the show, available at health food stores for the low, low price of only twenty dollars a bottle.)

Sadly, the "toxic teeth" scare has created a cottage industry for dentists willing to perform a procedure that is both unnecessary and dangerous—and one that, perhaps worst of all, provides false hope for those suffering from symptoms that have nothing to do with their mercury amalgams. Dentists who offer this service often advertise their practices with taglines such as "Chilling facts about the silent killer in your mouth."

It's not hard to scare people into action.

MERCURY amalgams are still available today, as are several other options. Patients can choose cast-gold fillings, which are durable, strong, and (to some) aesthetically pleasing. Like mercury amalgams, cast-gold fillings usually last ten to fifteen years. The problem with gold fillings is they cost about ten times more than mercury amalgams and require at least two office visits to put in place.

Patients could also choose composite resins, which match the color of teeth and are remarkably versatile; they are used not only to fill cavities but also to repair chipped, broken, or worn teeth. The problem with composite resins is that they wear out sooner than either gold fillings or mercury amalgams, often requiring replacement within five years. They also require more chair time for placement and are twice the price of mercury amalgams.

So, despite all the hype, in the end, the best choice might still be mercury amalgams.

PART VI

Common Beliefs

17

Vitamin C Doesn't Treat or Prevent Colds

Common colds are common for three reasons:

1. They're hard to prevent. The virus most likely to cause a cold is called rhinovirus (literally, "nose virus"); about one hundred different strains have been identified. Other viruses, such as influenza, parainfluenza, respiratory syncytial virus, and adenovirus, can also cause colds. Like rhinovirus, about one hundred different strains of adenovirus have also been described. All these viruses cause runny nose, congestion, sneezing, sore throat, headache, and fever. Because so many different strains of viruses cause the common cold, it's been impossible to make a vaccine to prevent it.

2. They're hard to treat. Antibiotics don't treat viruses. Although antiviral medications have been invented for some viruses, such as influenza and herpes, none has been invented to treat rhinoviruses.

3. They're easy to spread. Common cold viruses, which are spread primarily by sneezing, coughing, or even talking, can also be spread by

shaking hands or by touching a doorknob or an ATM machine or any other surface that an infected person recently touched.

For centuries, we've tried, largely in vain, to treat the common cold. We've taken cough suppressants, pain and fever medicines, antihistamines, expectorants, and natural products such as camphor, eucalyptus oil, zinc, and echinacea—all without much benefit and some, as in the case of antihistamines, that are dangerous for young children.

As a general rule, when so many different medicines are claimed to work, none of them probably does. Dr. William Osler, one of the founders of Johns Hopkins Hospital, lamented his profession's utter failure to deal with this disease. "The best way to treat the common cold," he said, "is with contempt."

THEN someone had a better idea: an "all-natural" product that he believed not only treated colds but prevented them.

Linus Pauling was born on February 28, 1901, in Portland, Oregon. In 1931, he published a scientific paper showing an alternative way for atoms to interact, marrying quantum physics with chemistry. It revolutionized the field. For this single paper, Pauling was named the most outstanding young chemist in the United States, became the youngest person ever elected to the National Academy of Sciences, was made a full professor at Caltech, and won the Nobel Prize in Chemistry. He was only thirty years old.

In 1949, Pauling published another paper, showing that the hemoglobin of patients with sickle cell anemia had a slightly different electrical charge—giving birth to the field of molecular biology.

In 1951, Pauling published a paper showing the unique way that some proteins folded upon themselves. He called one configuration the alpha-helix—later used by James Watson and Francis Crick to explain the structure of DNA. Again, Pauling was decades ahead of his peers.

In 1961, Pauling showed that mutations in hemoglobin could be used as a kind of evolutionary clock, proving that humans diverged from apes much sooner than scientists had suspected. A colleague later remarked, "At one stroke, he united the fields of paleontology, evolutionary biology, and molecular biology." That same year, Pauling appeared on the cover of *Time* magazine's "Men of the Year" issue, hailed as one of the greatest scientists who ever lived.

In the midst of these scientific accomplishments, Pauling took time to become the world's most recognized peace activist; his efforts led to the Nuclear Test Ban Treaty. As a consequence, in 1962, he won the Nobel Peace Prize, the first person ever to win two unshared Nobel Prizes.

Although Pauling's research never involved treating or preventing colds, that would soon change. By the early 1970s, the American public would come to know Linus Pauling for one thing and one thing only: vitamin C.

In March 1966, when he was sixty-five years old and giving a talk in New York City, Pauling said that he hoped he would live another twenty-five years. On his return to California, he received a letter from a man named Irwin Stone, who had been at the talk. Stone had spent two years studying chemistry in college, later receiving an honorary degree from the Los Angeles College of Chiropractic and a "PhD" from Donsbach University, a nonaccredited correspondence school. Stone wrote that if Pauling took 3,000 milligrams of vitamin C every day, he would live at least another twenty-five years.

Pauling followed Stone's advice. "I began to feel livelier and healthier," he recalled. "In particular, the severe colds I had suffered several times a year all my life no longer occurred. After a few years, I increased my intake of vitamin C to 10 times, then 20 times, and then 300 times the Recommended Dietary Allowance (RDA): now 18,000 milligrams per day!"

IN 1970, Pauling published *Vitamin C and the Common Cold,* urging the public to take six large 500-milligram tablets (3,000 milligrams) of vitamin C every day (about fifty times the RDA). Pauling's book was an instant bestseller. When the paperback edition hit the shelves, sales of vitamin C doubled, then tripled, then quadrupled. Drugstores couldn't keep up with the demand. Wholesale stocks were depleted. By the mid-1970s, 50 million Americans were following Pauling's advice. Vitamin manufacturers called it "the Linus Pauling Effect."

The only thing that Pauling's bestselling book lacked was scientific evidence proving that he was right—evidence that wasn't particularly hard to generate. All researchers had to do was divide volunteers into two groups. One group would receive vitamin C either at the time of illness or throughout the cough-and-cold season. The other group would receive a placebo pill that looked and tasted like vitamin C but wasn't. If vitamin C worked, then recipients would either have fewer colds or less severe colds, or both. Indeed, at the time Pauling published his book, two of these studies had already been done.

Vitamin C Doesn't Treat or Prevent Colds

In 1939, researchers at the University of Minnesota divided four hundred undergraduate students into two groups. One group was given 200 milligrams of vitamin C every day throughout the winter season, about twenty-eight weeks. Fifty of these students also took 500 milligrams of vitamin C during the first two days of a cold. The other group was given placebo tablets. Students were told to report to the student health center at the first sign of illness. When the winter season ended, the researchers found no differences between the two groups in the frequency, severity, or duration of colds.

Ten years later, these same investigators performed a similar study at the same school. This time, however, the dose of vitamin C was much greater. At the beginning of the 1948 school year, students were given 667 milligrams of vitamin C throughout the winter as well as an antihistamine. Other groups received an antihistamine alone, vitamin C alone, or placebo pills. Students were also asked to take 3,000 milligrams of vitamin C at the first sign of a cold and to continue taking it every day until symptoms were gone. Although the study design was different, the results were the same. The authors concluded that vitamin C didn't "have any important effect on the duration or severity of these infections."

Pauling refused to believe the Minnesota studies. In 1970, he wrote a letter to the *New York Times* decrying the investigators' conclusions. Pauling argued that the researchers had ignored positive results. He insisted that had they looked more closely they would have seen that some of these students had actually benefited from vitamin C. On December 26, 1970, the lead author on the Minnesota studies, Dr. Harold S. Diehl, the dean of medical sciences at the University of Minnesota, wrote a rebuttal letter that was also published in the *New York Times*. Diehl argued that Pauling had completely ignored critical data—in essence, that he didn't know what he was talking about.

Ironically, the same year that Linus Pauling criticized the Minnesota studies, the study to end all studies was performed by researchers at the University of Maryland School of Medicine on twenty-one prisoner volunteers at the House of Correction in Jessup, Maryland. First, blood specimens were taken from each of the volunteers and tested for immunity to a particular strain of a common cold virus called rhinovirus type 44. For two weeks, eleven of them were given 3,000 milligrams of vitamin C every day, and ten were given a placebo. Then all the men were inoculated intranasally with the strain of rhinovirus to which they were all susceptible. After the virus was inoculated, the vitamin C group continued to receive the vitamin for another week, and the placebo group continued to receive the placebo. Every day, the men were examined by physicians and their symptoms were recorded. And every day, nasal washings were obtained to determine the amount of rhinovirus that was excreted from the nose. As had been the case in the Minnesota trials, no differences were found in the frequency or severity of symptoms. In addition, no differences were found in the amount of rhinovirus that was shed during illness.

If you can get past the seemingly contradictory phrase "prisoner volunteers," in many ways, the Maryland study was perfect. Unlike trials of vitamin C in the general population, where people may be more or less susceptible to the strains of rhinovirus circulating in the community, the prisoner trial controlled for susceptibility to the challenge virus. Also, in the general population, people will be exposed to different quantities of virus, which will result in different severities of symptoms; the prisoner trial controlled for the dose of virus. Furthermore, whereas studies in the general population depend on self-reporting, the prisoner trial was supervised by physicians who examined the prisoners every day, providing a more accurate record of the frequency and severity of symptoms. The prisoner trial also determined the amount of rhinovirus that was shed from the nose in those who did or didn't receive vitamin C; no other studies before or since have done this. Finally, this trial controlled for the amount of vitamin C ingested; by definition, no one went off protocol, which often happens in studies of the general population where researchers depend on volunteers to do exactly what they are asked to do.

The prisoner study was so clear, so definitive, so well controlled, and so groundbreaking that Jane Brody, the highly quoted, highly acclaimed health reporter for the *New York Times*, wrote about it. Under the title "Vitamin C Study Rebuts Pauling," Brody described the details of the Maryland experiment. Again, Pauling refused to believe it, criticizing the study as "a poor investigation." He argued that the experiment was too small and that it was performed under unnatural conditions. It was becoming apparent that Linus Pauling possessed something that no scientist should ever possess: a nonfalsifiable belief.

Because he stood on the platform of two Nobel Prizes, Linus Pauling was an influential man. When he said that the studies performed in Minnesota and Maryland didn't make sense, people believed him. And so, the studies continued, all with the hope that someone somewhere would prove Linus Pauling right and that finally we would have something to treat or prevent a disease that has been untreatable and unpreventable.

In 1974, researchers at the University of Toronto divided 2,300 volunteers into seven groups. Three groups received either 250, 1,000, or 2,000 milligrams of vitamin C every day throughout the winter months; two groups received 4,000 or 8,000 milligrams at the start of illness; and two groups received placebo tablets. Again, researchers found no differences in the frequency or severity of infections.

One year later, in 1975, two independent groups of researchers reviewed fourteen more studies. One research group, using italics to emphasize their point, wrote that "a review of the controlled studies of the efficacy of [vitamin C] in the prophylaxis and therapy of the common cold . . . reveals *little convincing evidence to support claims of clinically important efficacy*." The other group, noting that not all studies were well performed, stated that "the minor benefits of questionable validity are not worth the potential risk [of vitamin C], no matter how small that might be."

Still, the studies continued.

In 1977, researchers in Australia divided one thousand subjects into two groups. One group was given 1,000 milligrams of vitamin C every day during the winter season and 3,000 milligrams at the beginning of illness for three days. The other group was given placebo tablets. Again, no difference. The researchers concluded, "It is becoming clear from this and other work that the effect of [vitamin C] on the common cold is at best elusive and probably trivial."

Twenty years later, in 1997, researchers reviewed the results of six more studies of vitamin C, involving more than five thousand episodes of illness, concluding that "it is shown that the common cold incidence is not reduced in the vitamin C supplemental groups compared with placebo groups."

Nonetheless, in what apparently had become an endless quest to prove Linus Pauling right, the studies continued.

In 2001, researchers at the Australian National University in

Canberra divided four hundred healthy volunteers into four groups; one group received only 30 milligrams of vitamin C at the onset of colds, to be taken for two days (the placebo group); two groups received either 1,000 or 3,000 milligrams; the fourth group received 3,000 milligrams plus a popular product called Bio-C, which contained a variety of natural substances such as rose hip extract and bioflavonoids. Again, the authors found no differences in the frequency, duration, or severity of cold symptoms.

Finally, in 2013, researchers reviewed the results of twenty-nine more studies of vitamin C, involving more than eleven thousand participants, concluding, "The failure of vitamin C supplementation to reduce the incidence of colds in the general population indicates that routine vitamin C supplementation is not justified."

Because the evidence is clear, the Food and Drug Administration, the American Academy of Pediatrics, the American Medical Association, the Academy of Nutrition and Dietetics, the Center for Human Nutrition at Johns Hopkins Hospital, and the Department of Health and Human Services do not recommend supplemental vitamin C for the prevention or treatment of colds.

Linus Pauling never wavered. Up until the day he died, on August 19, 1994, he continued to promote vitamin C in speeches, popular articles, and books, refusing to believe that he could ever be wrong. When he appeared before the media with cold symptoms, he said he was suffering from allergies.

IN medicine, if something works to treat or prevent an illness, it's been relatively easy to prove it. Insulin treats diabetes. Chemotherapies treat certain cancers. Vaccines prevent viral and bacterial infections. Researchers in these situations didn't have to perform dozens of studies before they showed that their interventions worked. So, given that vitamin C has consistently been proved to be useless in treating or preventing colds, why does it remain so popular?

The answer shouldn't be surprising.

In February 2012, the pharmaceutical giant Pfizer bought Alacer

Corporation, the makers of a popular vitamin C product called Emergen-C. "We are very pleased that the Emergen-C family of products will become part of Pfizer's portfolio," said Paul Sturman, president of Pfizer Consumer Healthcare. Although financial terms of the purchase weren't disclosed, it's not hard to understand why Sturman was excited. At the time of the purchase, Alacer was producing about 500 million packets of Emergen-C a year, with revenues exceeding $220 million. Emergen-C was sold in health food stores, supermarkets, drugstores, and campus bookstores. And Alacer was only one of many vitamin C manufacturers.

We continue to take vitamin C for our colds because manufacturers spend millions of dollars every year convincing us that it's the right thing to do. Medical and scientific journals, on the other hand, don't financially benefit from the results of their studies; not surprisingly, these journals don't spend money advertising their findings.

Also, apart from the fact that vitamin C isn't free, the downside of taking something that clearly doesn't work is that, as is true for any medication, it can have side effects. According to the Mayo Clinic, large quantities of supplemental vitamin C can cause headache, nausea, vomiting, dizziness, inflammation of the esophagus, flushing, and fatigue.

THERE is, however, one treatment for the common cold that really does work. And unlike supplemental vitamin C, it's cheap and harmless. It's also been recommended by mothers for decades: chicken soup. But it's not the chicken or the broth or the celery or the carrots or the noodles or the rice that makes a difference; it's the steam.

Researchers have always known that rhinoviruses survive better at the colder temperatures found in the nose (91.4 to 95 degrees) than at core body temperature (98.6 degrees). In 2015, in a paper published in the prestigious *Proceedings of the National Academy of Sciences*, researchers at Yale University School of Medicine reported on their investigation of this phenomenon. They took airway cells from mice and exposed them to a strain of rhinovirus at lower temperatures and higher temperatures. As expected, rhinovirus couldn't reproduce at the higher temperatures. Then they found out why. The higher temperatures triggered airway cells to produce an antiviral substance called interferon, which limited the ability of rhinoviruses to replicate.

As it turns out, after all these years, a treatment for the common cold that offers some benefit has been right under our nose.

18

Don't Ice Sprains

A basketball player leaps up, grabs a rebound, comes down on another player's foot, twists his ankle, and screams in pain.

Basketball, squash, tennis, ice-skating, volleyball, and soccer all have one thing in common: explosive movements. For this reason, almost everyone who plays these sports will at one time or another sprain their ankle. Indeed, ankle sprains account for 20 percent of *all* sports injuries. But you don't have to be an athlete to sprain your ankle. People sprain their ankles when they step awkwardly off a curb or slip on a wet patch. Every day, twenty thousand people in the United States sprain their ankle, and every day they put ice on the sprain. What many people fail to realize is that this is exactly the wrong thing to do.

WHEN people sprain their ankle, they injure three ligaments that stabilize the outside of the joint. Ligaments are short, tough, flexible, fibrous bands that connect one bone to another and hold the joint together. The seriousness of an ankle injury is determined by how badly these ligaments are damaged. Doctors assign three grades:

Grade 1: Ligaments are stretched but not torn. People with grade 1 ankle injuries have no loss of function and suffer only minimal pain, swelling, and difficulty bearing weight.

Grade 2: Ligaments are partially torn, causing some looseness of the joint. People with grade 2 ankle injuries suffer some loss of function as well as moderate pain, swelling, and difficulty bearing weight.

Grade 3: Ligaments are completely torn. People with grade 3 ankle injuries suffer complete loss of function as well as severe pain, swelling, and an inability to bear any weight. Grade 3 sprains also cause bleeding in the joint.

If ligaments in the joint don't heal completely, long-term ankle instability can result. Indeed, up to 30 percent of people with severe ankle sprains suffer permanent looseness of the joint. For this reason, those first few days of treatment are crucial.

ALTHOUGH many have offered a variety of treatments for ankle sprains, in the late 1970s, one physician, Gabe Mirkin, revolutionized the field.

Mirkin is a man of many talents. After graduating from Harvard University and Baylor College of Medicine, he developed an expertise in allergy, immunology, sports medicine, and pediatrics. During one of his fellowships, specifically at Johns Hopkins Hospital in Baltimore, Mirkin began training for long-distance races, running twice a day. When his fellowship ended, he moved to Washington, DC, because it was the only city in the United States that offered competitive racing fifty-two weeks a year.

In 1978, when Dr. Mirkin was forty-three, he had his breakthrough moment, writing an instant bestseller titled *The Sports Medicine Book*. Although Mirkin discussed many sports-related injuries, it was his advice on how to treat sprains that became a mantra for orthopedists, sports medicine doctors, and team trainers. He called it RICE, an acronym for rest, ice, compression, and elevation. For decades, RICE became the go-to treatment for groups such as the American Academy of Family Physicians, the American Academy of Orthopedic Surgeons, and the National Athletic Trainers' Association. Unfortunately, Mirkin's advice was based more on instinct than evidence.

The first cracks in the wall appeared in the late 1980s.

Icing Sprains Delays Healing

In 1989, researchers from New Zealand divided thirty adult patients with ankle sprains into two groups. One group received an ice pack for twenty minutes; the other didn't. The authors concluded that "no significant differences were noted between the two groups with regard to pain, swelling or speed of recovery."

In 2006, researchers from Ireland divided eighty-nine athletes and nonathletes with grade 2 or 3 ankle sprains into two groups. Patients received either ten minutes or twenty minutes of ice therapy every two hours for three days. At the end of the experiment, patients with *less* ice therapy had less pain.

In 2012, researchers from Amsterdam reviewed eleven more studies, involving 870 patients, to determine whether RICE improved outcomes. Again, they couldn't find evidence to support Mirkin's recommendations. In addition to the lack of effectiveness for ice, the authors couldn't find evidence that rest, compression, or elevation worked, either. Indeed, they found three studies that showed that early movement instead of rest led to better outcomes.

In 2013, researchers from Taiwan performed a study that explained why ice worsened outcomes. Eleven male subjects, all twenty years of age, performed six sets of elbow extensions at 85 percent maximum tolerance. This level of exercise is enormously strenuous. When the exercise concluded, researchers measured the subjects' blood for two proteins that are found in muscle cells: creatine kinase and myoglobin. High levels of these proteins had spilled into the bloodstreams of all these young men, proving that the exercise had caused some minor muscle damage.

The researchers then divided the subjects into two groups. In one group, cold packs were applied to the stressed muscles for fifteen minutes immediately after the exercise as well as three, twenty-four, forty-eight, and seventy-two hours later. The other group received no ice therapy. The researchers found that the ice therapy group had significantly higher levels of the muscle proteins, and therefore more muscle damage, than the nontreatment group. Consistent with these findings, seventy-two hours after the exercise ended, the ice group was more fatigued than the no-ice group.

Why Does Icing Worsen Outcomes?

The key to healing is inflammation. When ligaments or muscles are damaged, they release substances that promote inflammation. In response, the body increases blood flow to the area. The bad news is that inflammation is painful. The good news is that it promotes healing. Here's how:

- Increased blood flow brings clotting factors to the damaged area, which helps stop the small amount of bleeding that inevitably occurs when ligaments or muscles are damaged.
- Increased blood flow brings immune cells to the area, which helps remove damaged ligament and muscle cells.
- Increased blood flow promotes the manufacture of collagen, which is necessary to create new, healthy ligaments and muscles.

Therefore, anything that decreases blood flow (such as rest, ice, compression, and elevation) or decreases inflammation (such as ibuprofen) only lengthens the time to healing.

In 2013, about thirty years after publication of *The Sports Medicine Book*, Dr. Mirkin, to his credit, recanted. "There are no data to show that ice does anything more than block pain," he said. "And there are data that show it delays healing. RICE is just something that stuck—and it's wrong." Regarding rest, Mirkin said, "Nobody believes in rest anymore. You can get a hip replacement and you're on the bike 12 hours after surgery." As a consequence of recent studies, as well as Dr. Mirkin's mea culpa, a small but growing number of doctors no longer recommend RICE for joint or muscle strains.

SO, what are you supposed to do now? The simple truth is that sprains and strains hurt and that decreasing inflammation, whether with rest, ice, compression, elevation, or anti-inflammatory drugs, lessens the pain. Anything that lessens pain will be hard to resist. Unfortunately, we now know that RICE and anti-inflammatory drugs delay healing. The better choice would be to *warm* the affected joint or muscle. Regarding ankle sprains, the current recommendation isn't complete rest but, rather, some movement; specifically, don't bear weight if it still hurts, but pretend to write letters of the alphabet with your toes within a couple of days of the sprain. This gentle continuous movement promotes blood flow to the area, and therefore healing.

The situation of treating the pain of sprains or strains with ice is analogous to treating fever caused by infections. Although we feel better when the fever is gone, we might prolong or worsen illness. The process of inflammation, although painful and uncomfortable, is necessary for recovery. We pay a price when we try to eliminate it.

19

Teething Doesn't Cause Fever

The jokes say it all:

> "The Tooth Fairy's evil twin is the Teething Fairy" (popular meme).
>
> "I heard whiskey helps with teething, so I drank a shot before bed last night. Turns out it's true" (another popular meme).
>
> "Adam and Eve had many advantages, but the principal one was that they escaped teething" (Mark Twain).
>
> "Watching teething babies is like watching over a thermonuclear reactor—it is best done in shifts, by well-rested people" (Anthony Doerr).

Parents have been talking about teething for at least three thousand years. Hippocrates claimed that "teething children suffer from itching gums, fever, convulsions, and diarrhea, especially when they cut their eye teeth [incisors] and are corpulent [overweight] or costive [constipated]." In 1839, more than five thousand infant deaths in England and Wales were attributed to teething. Not everyone, however, agreed. The first to argue that teething didn't cause an array of potentially fatal symptoms was François Ranchin, a seventeenth-century professor in southern France. Centuries later, Ronald Illingworth, a pediatrician and author of the book *The Normal Child,* offered the best summary of current thinking: "Teething produces nothing but teeth," he wrote. Nevertheless, virtually all parents today believe that, at the very least, teething causes fever. Why?

Teething is a natural process that involves the tooth traveling up or down from the jaw, breaking through the gums, and entering the mouth. It begins between four and ten months of age with the lower central incisor. By the time the child is twenty-four months old, all twenty primary teeth have usually appeared.

In defense of popular belief, it's reasonable to assume that a process that involves disruption of the gums could cause inflammation and fever. However, it's also not hard to prove whether it really does. All you need to do is examine infants every day, check their gums to determine when primary teeth have broken through, and record temperatures and symptoms. These studies have been done, and the results have always been the same.

Teething Doesn't Cause Fever

In 2000, Australian investigators studied twenty-one children between six and twenty-four months of age who attended several suburban day care centers. Every morning for seven months, the investigators took the children's temperature, and every morning they examined the children's gums. When the study concluded, the investigators compared 236 teething days with 895 non-teething days. The average temperature during teething days was 97.18 degrees and during non-teething days was 97.12 degrees. The investigators also found no differences in the presence of other symptoms, concluding, "This study did not confirm that tooth eruption in infants and toddlers is associated with fever, mood disturbance, an appearance of illness, sleep disturbance, drooling, diarrhea, strong urine, red cheeks, or rashes/flushing on the face or body."

In 2011, Brazilian investigators studied forty-seven infants between five and fifteen months of age. As in the Australian study, every day for eight months the investigators examined each child's gums for tooth eruption, and every day they took each child's temperature. During the study, 231 teeth erupted, with each child averaging about five new teeth. The maximal temperature during teething was 98.2 degrees.

In 2017, Brazilian researchers reviewed the results of six more studies of fever during teething and concluded, "In the overall meta-analysis, no association was found between fever and primary tooth eruption."

If teething doesn't cause fever, then why do so many people believe that it does? The answer lies in *when* primary teeth erupt. While the baby is in the womb, it receives its blood supply from the mother, which includes antibodies directed against a variety of bacteria and viruses. At the time of birth, babies are equipped with these antibodies. The half-life of these maternal antibodies (meaning the time in which half can no longer be detected) is only about twenty-one days; therefore, by six months of age, the protection afforded by these antibodies is gone. Also, many women will have stopped breastfeeding by the end of the child's first year of life, which eliminates additional antibodies that bathe the child's throat and intestines. For these reasons, viral diseases of the ear, nose, throat, and intestine occur most commonly between six and twenty-four months of age—the same time that children are teething. This explains why parents often associate teething with symptoms such as fever, runny nose, and diarrhea.

Unfortunately, scapegoating teething as a cause of fever can be harmful. If a child has a fever, especially a high fever, parents should make sure that they aren't ignoring a potentially treatable infection.

Epilogue: Choosing Wisely

How did we get here? How did we get to a place where doctors occasionally ignore thoughtful, carefully performed studies published in excellent journals and order tests or prescribe drugs or perform cancer screenings that have clearly been shown to do more harm than good? This problem isn't exactly a well-kept secret. In 2017, *The Atlantic* published an article titled "When Evidence Says No, but Doctors Say Yes." One year later, an op-ed in the *New York Times* read, "It's Hard for Doctors to Unlearn Things: That's Costly for All of Us."

So, again, how did we get here?

Ignorance

Sometimes doctors simply aren't up to date on recent studies. Given the complexity and diversity of medicines and therapies, it's not hard to see how this could happen. There are 6,500 medical and scientific journals in the world that publish about 4,000 studies every day. Not surprisingly, these articles follow a bell-shaped curve. Some are excellent. Some are awful. Most are more or less mediocre. It can be difficult to separate the good ones from the bad ones, especially for a busy clinician.

Fee for Service

Doctors are paid for what they do, not for what they don't do. "It's hard to get people in any profession to do the right thing," wrote David Epstein for the public interest group ProPublica, "when they're paid to do the wrong thing." This is particularly true for surgical procedures such as knee arthroscopies and heart stents.

Steve Nissen, chairman of cardiovascular medicine at the Cleveland Clinic, offered one solution: eliminate financial incentives. "Things have gotten better but they're not where they need to be," he said. "I have a dozen or so cardiologists, and they get the exact same salary whether they put in a stent or don't. I think that's made a difference and kept our rates of unnecessary procedures low."

Regarding inappropriate testing, such as measuring blood levels of testosterone and vitamin D, Canada and Australia simply stopped paying for them. Within six months, the number of tests ordered dropped by 80 percent.

Patient Satisfaction

In the United States, health care is a $3.3 trillion business. Unfortunately, it often acts like a business. Patients aren't patients anymore; they're clients. Customer satisfaction has become the coin of the realm.

Dr. John Christoforetti, a prominent orthopedic surgeon in Pittsburgh, shared an experience that is both sobering and typical. After evaluating a woman who wanted an operation to repair a small meniscal tear in her knee, Christoforetti explained to her and her husband that she didn't need it. Physical therapy would work just as well. "She left with a smile on her face," Christoforetti recalled, "but literally as they were checking out we got a ding that someone had rated us on a website and it's her husband. He's been typing on his phone during the visit, and it's a one-star rating that I'm an insensitive guy he wouldn't let operate on his dog. They'd been online and they firmly believed she needed this one operation and that I was the guy to do it. Most of my colleagues will say, 'Look, save yourself the headache and just do the surgery. Your bank account's not going to be upset with you for doing the surgery. Just do the surgery.'"

Demand for unnecessary antibiotics is another example. In 2018, researchers from the Cleveland Clinic examined the relationship between patient ratings and receipt of an antibiotic prescription. The study included adults with colds and mild upper respiratory infections, the vast majority of which are caused by viruses and don't benefit from antibiotics. The researchers examined 8,500 encounters with 85 physicians. They found that 91 percent of patients who received an antibiotic prescription rated their encounter with five stars compared with 73 percent of those who didn't receive a prescription. To reach the top twenty-fifth percentile in patient satisfaction, physicians had to prescribe unnecessary antibiotics at least half the time; to reach the top tenth percentile, they had to prescribe them at least 75 percent of the time. The authors concluded, "No other patient or physician factor was as strongly associated with patient satisfaction as receipt of a prescription for an antibiotic."

Unfortunately, patient satisfaction scores often conflict with best medical practices. Doctors in hospitals are starting to act more like waiters in restaurants. ("Hi. My name is Dr. Smith. I'll be your doctor today. Let me tell you about a few of the therapies that we have on our menu.") Shannon Brownlee from the Lown Institute, an organization dedicated to restoring the doctor-patient relationship, wrote, "Calling patients consumers who need to be engaged reinforces an attitude that says that a person who is sick, frightened, in pain, and in need of care is no different than a consumer buying a used car."

Health care providers can best guide patients through the dense thicket of medical information by offering booklets, videos, tear sheets, and other educational materials instead of caving in to patient demands. Several physician groups, such as Choosing Wisely, the Right Care Alliance, and Slow Medicine, are now doing just that, offering easy-to-understand educational materials so patients can make the best choice for their health, even when it's not the easiest or most popular choice.

Inertia

Most physicians take pride in what they do. It's hard for them to imagine that they've been doing something for years that might have been hurting their patients. Even in the face of overwhelming evidence to the contrary, established practices die hard.

In 2007, John Ioannidis and colleagues, in a paper published in the *Journal of the American Medical Association* titled "Persistence of Contradicted Claims in the Medical Literature," found that even after particular claims had been dismissed by excellent studies, many publications continued to reference these older, unsupportable claims. For example, they examined the claims that vitamin E could decrease the incidence of heart disease; that beta-carotene, a vitamin A precursor, could prevent cancer; and that estrogen treated Alzheimer's disease—all of which had been shown to be false. They found that it took about ten years before medical publications acknowledged that these claims were invalid. The authors argued that the reason it took so long was that doctors wanted the original claim to be true. They wanted to believe that they had been helping. "Sometimes investigator beliefs in scientific circles may have similar psychological characteristics as the nonscientific beliefs observed in other areas of society," they wrote. "The wish bias of individuals irrespective of topic can be large."

Perhaps the best example of the physician's desire to hold on to the belief that a disproven therapy is valuable is heart stents. In 2018, researchers in the United Kingdom randomly assigned patients with at least 70 percent blockage of a single coronary artery into two groups. One group received a stent to open up the blockage; the other underwent a sham operation. Called ORBITA (Objective

Randomised Blinded Investigation with Optimal Medical Therapy of Angioplasty in Stable Angina), it was the perfect study. After the real or sham operations, both groups improved. Merely believing that you had had a stent placed improved exercise tolerance. Although doctors could reasonably argue about the ethics of this study, they couldn't argue with the results. Nevertheless, they did.

After publication of this trial, heart surgeons railed against the result. After all, heart stents made biological sense: a patient has chest pain; a doctor finds a blockage; removing the blockage should relieve the pain. As it turns out, it was more complicated than that. Blockage of a large coronary artery was also invariably accompanied by blockages of smaller branches downstream, which weren't treated by the stent. Doctors wrote letters to journals and posted comments about the trial on social media. They claimed that the patients in the trial weren't sick enough; or that a subgroup might have benefited; or that, in their experience, stents worked; or that the trial was unethical because everyone already knew that stents worked. One of the principal investigators on the trial, Darrel Francis, a British interventional cardiologist, wrote on Twitter, "I would like to thank the patients participating in #ORBITA without realizing what a low level of intellect would be put into criticizing their efforts."

In the 1950s, Dr. Paul Hawley, director of the American College of Surgeons, said, "[T]he public would be shocked if it knew the amount of unnecessary surgery performed." In 1976, the American Medical Association called for a congressional hearing on unnecessary surgeries, claiming that every year, "2.4 million unnecessary operations [were] performed on Americans at a cost of $3.9 billion." Unfortunately, surgical procedures are not highly regulated. Unlike drugs, surgeries aren't subjected by the Food and Drug Administration to Phase I, II, and III studies where researchers have to prove in progressively larger clinical trials that their therapies work and are safe. We are well beyond the time when it would be of value to impose these regulations.

Medical Marketing

Between 1997 and 2016, spending on the marketing of drugs, disease awareness campaigns, health services, and laboratory testing increased from $12.7 billion to $29.9 billion a year. Direct-to-consumer ads alone increased from $2.1 billion to $9.6 billion. Disease awareness campaigns for bogus disorders such as "Low T" offer catchy, easy-to-remember acronyms (e.g., ED, for erectile dysfunction), quizzes that encourage self-diagnosis, and requests for patients to "ask their doctor" by providing scripted questions about treatment. Scientific studies showing that patients are being duped don't stand a chance against this onslaught of misleading information. (Of interest, only the United States and New Zealand allow direct-to-consumer advertising of drugs.)

Worse, megavitamins and dietary supplements, apart from marketing budgets approaching $1 billion, aren't effectively regulated. The FDA fails to counter bogus claims that dietary supplements support joint health or heart health or prostate health or that they boost the immune system or increase energy or reduce stress, because the powerful dietary supplement industry has influenced legislation, effectively tying the FDA's hands. Again, studies showing that vitamin D doesn't treat a variety of illnesses can't possibly compete with a misleading marketing barrage claiming that it does.

To underline the power of medical marketing, a pharmacist at Children's Hospital of Philadelphia told the following story: "As a pharmacy student thirteen to fourteen years ago, I had a four-week rotation at the VA Hospital in Huntington, West Virginia. I had a woman come into the clinic and, after her blood pressure check, she said, 'I need to talk to the doctor today, too.' I asked, 'What seems to be bothering you?' In a whispered voice she said, 'I think I have a "going" problem and would like to talk to the doctor about it.' She wanted the doctor to prescribe Avodart or Flomax based on the commercials she had seen. These are medications for men with enlarged prostates. I explained to her that her problem wouldn't be treated the same way [given that she didn't have a prostate], but she remained unconvinced. She wanted those medicines." Such is the power of drug advertising.

Fear of Lawsuits

Many physicians practice what they call "defensive medicine." The concern is that they are more likely to get sued for what they don't do than for what they do. For example, if a patient receives a heart stent and dies following the procedure, it would be difficult for the patient to sue and win. Surgeries have risks, and people occasionally die following the placement of stents—as is clearly stated on the consent-for-surgery form. However, if the patient is found to have blockage of a major coronary artery and a stent isn't placed, the doctor might be sued successfully if the patient then suffers a heart attack and dies—even if physical therapy plus medication would have been the better, safer choice.

The solution to this problem is simple: eliminate jury trials. If a patient or family feels that a doctor has engaged in medical malpractice (negligence resulting in harm), then the case should be presented to a commission composed of experts in the field who can most fairly judge whether it really was a case of malpractice. Medical and scientific issues can't be accurately judged by people without expertise and experience in the subject.

Although the following case takes this problem to its illogical end, it shows just how ridiculous some jury verdicts can be.

In 1986, Judith Haimes, a self-proclaimed psychic, had a headache. She made an appointment to see a neurologist at Temple University in Philadelphia, who ordered a CT scan of her head. The scan was normal. Unfortunately, at least according to Haimes, her psychic powers had now vanished. No longer would she be able to make a living by communicating with the dead. Haimes sued Temple and the neurologist. On March 27, 1986, after deliberating for forty-five minutes, the eight-member jury awarded her $986,000. In order for jury members to have reached this verdict, they would have to have agreed on three separate facts: one, that people have psychic powers to lose; two, that psychic powers can be lost in a CT scan; and three, that the neurologist should have known better than to have ordered a CT scan on a psychic ("What was he thinking?"). Probably the only chance that the defense had to win this case, given the belief systems of the jury, would have been to argue that Haimes should have seen it coming. In August 1986, an appeals court judge overturned the verdict, arguing that it was "so grossly excessive as to shock the court's sense of justice."

Unfortunately, given today's litigious climate and the influence of the plaintiff's bar, eliminating nonmedical juries has about as much chance of succeeding as Judith Haimes has of communicating with the dead.

PEOPLE shouldn't be expected to be their own doctors, but they should be informed about their health so that at the very least, they can ask good questions. If your doctor prescribes vitamin D or recommends a heart stent or a thyroid screen or a prolonged course of antibiotics, you would do well to know the facts. Then and only then can you be the best advocate for your health and your children's health. Otherwise, you will continue to fall prey to the inertia, financial influences, and dogma that continue to allow for outmoded, unnecessary, and potentially dangerous therapies.

Acknowledgments

I would like to thank Dara Kaye and Gail Ross for their support of this project and the following colleagues, family members, collaborators, and friends for their careful reading of the manuscript and thoughtful suggestions: Hamid Bassiri, Louis Bell, Heather Bodenstab, Susan Coffin, Brian Fisher, Sarah Geoghegan, Jeff Gerber, Lori Kestenbaum Handy, Joyce Lee, Charlotte Moser, Bonnie Offit, Carl Offit, Julia Sammons, Michael Silberman, Sanjeev Swami, Anne Titterton, and Laura Vella.

Notes

Prologue: Ignoring James Lind

3 James Lind died on July 13: P. M. Dunn, "James Lind (1716–94) of Edinburgh and the Treatment of Scurvy," *Archives of Disease in Childhood* 76 (1997): F64–F65; M. Bartholomew, "James Lind's 'Treatise of the Scurvy' (1753)," *Postgraduate Medicine* 78 (2002): 695–96; J. H. Baron, "Sailors' Scurvy Before and After James Lind—A Reassessment," *Nutrition Reviews* 67 (2009): 315–32; I. Milne, "Who Was James Lind, and What Exactly Did He Achieve?" *Journal of the Royal Society of Medicine* 105 (2012): 503–8; Anonymous, "James Lind and Scurvy: The First Clinical Trial in History," *OpenMind*, July 12, 2016; M. White, "James Lind: The Man Who Helped to Cure Scurvy with Lemons," BBC News, October 4, 2016.

Chapter 1: Treating Fever Can Prolong or Worsen Illness

9 malaria therapy for syphilis: G. Gartlehner and K. Stepper, "Julius Wagner-Jauregg: Pyrotherapy, Simultanmethode, and 'Racial Hygiene,'" *Journal of the Royal Society of Medicine* 105 (2012): 357–59; C. J. Tsay, "Julius Wagner-Jauregg and the Legacy of Malarial Therapy for the Treatment of General Paresis of the Insane," *Yale Journal of Biology and Medicine* 86 (2013): 245–54.

10 Hippocrates believed: M. J. Kluger, "Fever in Ectotherms: Evolutionary Implications," *American Zoology* 19 (1979): 295–304.

10 injected 140 lizards: M. J. Kluger, D. H. Ringler, and M. R. Anver, "Fever and Survival," *Science* 188 (1975): 166–68; H. A. Bernheim and

M. J. Kluger, "Fever: Effect of Drug-Induced Antipyresis on Survival," *Science* 193 (1976): 237–39.

11 animals . . . treated or not treated with antipyretics: M. J. Kluger, "Fever Revisited," *Pediatrics* 90 (1992): 720–24; J. B. Covert and W. W. Reynolds, "Survival Value of Fever in Fish," *Nature* 267 (1977): 43–45; Kluger, "Fever in Ectotherms"; H. V. Ellingson and P. F. Clark, "The Influence of Artificial Fever on Mechanisms of Resistance," *Journal of Immunology* 43 (1942): 65–83; A. Lwoff, "Factors Influencing the Evolution of Viral Diseases at the Cellular Level and the Organism," *Bacterial Reviews* 23 (1959): 109–24; D. Rodbard, "The Role of Regional Body Temperature in the Pathogenesis of Disease," *New England Journal of Medicine* 305 (1981): 808–14; E. Atkins, "Treatment of Fever," *New England Journal of Medicine* 309 (1983): 925.

11 University of Illinois . . . rhinovirus: E. D. Stanley, G. G. Jackson, C. Panusarn, et al., "Increased Virus Shedding with Aspirin Treatment of Rhinovirus Infection," *Journal of the American Medical Association* 231 (1975): 1248–51.

11 Johns Hopkins . . . chicken pox: T. F. Doran, C. De Angelis, R. A. Baumgardner, et al., "Acetaminophen: More Harm Than Good for Chickenpox," *Journal of Pediatrics* 114 (1989): 1045–48.

11 University of Adelaide . . . rhinovirus: N. M. H. Graham, C. J. Burrell, R. M. Douglas, et al., "Adverse Effects of Aspirin, Acetaminophen, and Ibuprofen on Immune Function, Viral Shedding, and Clinical Status in Rhinovirus-Infected Volunteers," *Journal of Infectious Diseases* 162 (1990): 1277–82.

12 University of Helsinki . . . salmonella: A. S. El-Radhi, T. Rostila, and T. Vesikari, "Association of High Fever and Short Bacterial Excretion After Salmonellosis," *Archives of Diseases of Childhood* 67 (1992): 531–32.

12 Fujimoto Children's Hospital . . . viral infections: T. Sugimura, T. Fujimoto, H. Motoyama, et al., "Risks of Antipyretics in Young Children with Fever Due to Infectious Disease," *Acta Paediatrica Japonica* 36 (1994): 375–78.

12 University of Maryland . . . influenza virus or shigella: K. I. Plaisance, S. Kudaravalli, S. S. Wasserman, et al., "Effect of Antipyretic Therapy on the Duration of Illness in Experimental Influenza, *Shi-*

gella sonnei, and *Rickettsia rickettsii* Infections," *Pharmacotherapy* 20 (2000): 1417–22.

12 University of Miami . . . intensive care unit: C. I. Schulman, N. Namias, J. Doherty, et al., "The Effect of Antipyretic Therapy Upon Outcomes in Critically Ill Patients: A Randomized, Prospective Study," *Surgical Infections* 6 (2005): 369–75.

12 Hungary . . . sepsis: Z. Rumbus, R. Matics, P. Hegyi, et al., "Fever Is Associated with Reduced Hypothermia with Increased Mortality in Septic Patients: A Meta-Analysis of Clinical Trials," *PLOS One* (2017), doi: 10.1371/journal.pone.0170152.

13 primitive and adaptive: S. S. Evans, E. A. Repasky, and D. T. Fisher, "Fever and the Thermal Regulation of Immunity: The Immune System Feels the Heat," *Nature Reviews* 15 (2015): 345–49.

15 Czech Republic . . . vaccines: R. Prymula, C.-A. Siegrist, R. Chilbek, et al., "Effect of Prophylactic Paracetamol Administration at Time of Vaccination on Febrile Reactions and Antibody Responses in Children: Two Open-Label, Randomised Controlled Trials," *Lancet* 374 (2009): 1339–50.

15 Australia . . . influenza vaccine: J. Li-Kim-Moy, N. Wood, C. Jones, et al., "Impact of Fever and Antipyretic Use on Influenza Vaccine Immune Responses in Children," *Pediatric Infectious Disease Journal*, February 19, 2018, doi: 10.1097/INF.0000000000001949.

15 antipyretics during an influenza epidemic: J. D. Earn, P. W. Andrews, and B. M. Bolker, "Population-Level Effects of Suppressing Fever," *Proceedings of the Royal Society* 281 (2014): 1–5.

15 plants . . . increase their temperatures: C. E. Yarwood, "Heat of Respiration of Injured and Diseased Leaves," *Phylopathology* 43 (1953): 675–81.

16 heatstroke: A. Bouchama and J. P. Knochel, "Heat Stroke," *New England Journal of Medicine* (2002) 346: 1978–88.

17 suffer a "febrile seizure": M. Offringa and R. Newton, "Prophylactic Drug Management for Febrile Seizures in Children," *Cochrane Database Systematic Reviews* (2012), doi: 10.1002/14651858.CD003031.pub2; D. S. Seinfeld and J. M. Pellock, "Recent Research on Febrile Seizures: A Review," *Journal of Neurology and Neurophysiology* (2013), doi: 10.4172/2155-9562.1000165.

Chapter 2: Finishing the Antibiotic Course Is Often Unnecessary

21 Tom Patterson was dying: Steffanie Strathdee and Thomas Patterson, *The Perfect Predator: A Scientist's Race to Save Her Husband from a Deadly Superbug* (New York: Hachette Books, 2019).

22 design and test bacteriophages: K. Servick, "U.S. Center Will Fight Infections with Viruses," *Science* 360 (2018): 1280–81.

23 "Antibiotic Awareness Week": M. J. Llewelyn, J. M. Fitzpatrick, E. Darwin, et al., "The Antibiotic Course Has Had Its Day," *British Medical Journal* (2017), doi: 10.1136/bmj.j3418.

23 bacteria will become resistant: C. Costelloe, C. Metcalfe, A. Lovering, et al., "Effect of Antibiotic Prescribing in Primary Care on Antimicrobial Resistance in Individual Patients: Systematic Review and Meta-Analysis," *British Medical Journal* 358 (2010): c2096, doi: 10.1136/bmj.c2096.

23 risk of kidney stones: G. E. Tasian, T. Jemielita, D. S. Goldfarb, et al., "Oral Antibiotic Exposure and Kidney Stone Disease," *Journal of the American Society of Nephrology* 29 (2018), doi: 10.1681/ASN .2017111213.

24 their first patient, Albert Alexander: Llewelyn, Fitzpatrick, Darwin, et al., "The Antibiotic Course Has Had Its Day."

25 average life expectancy: E. Martens and A. L. Demain, "The Antibiotic Resistance Crisis, with a Focus on the United States," *Journal of Antibiotics* 70 (2017): 520–26.

25 highly resistant to antibiotics: Martens and Demain, "The Antibiotic Resistance Crisis."

25 "a post-antibiotic era": C. Nathan and O. Cars, "Antibiotic Resistance—Problems, Progress, and Prospects," *New England Journal of Medicine* 371 (2014): 1761–63.

25 WHO now estimates: S. LaFee and H. Buschman, "Novel Phage Therapy Saves Patient with Multidrug-Resistant Bacterial Infection," April 25, 2017, https://health.ucsd.edu/news/releases/Pages/2017-04-25-novel-phage-therapy-saves-patient-with-multidrug-resistant-bacterial-infection.aspx.

26 problem of MRSA: Martens and Demain, "The Antibiotic Resistance Crisis," 520–26.

26 The story of MRSA: F. D. Lowy, "Antimicrobial Resistance: The Example of *Staphylococcus Aureus*," *Journal of Clinical Investigation* (2003), https://www.jci.org/articles/view/18535.

26 Carl Nathan . . . "golden era": Nathan and Cars, "Antibiotic Resistance," 1761–63.

26 difficult to identify: S. E. Cosgrove and Y. Carmeli, "The Impact of Antimicrobial Resistance on Health and Economic Outcomes," *Clinical Infectious Diseases* 36 (2003): 1433–37.

27 more dangerous than those: Cosgrove and Carmeli, "The Impact of Antimicrobial Resistance."

27 30–50 percent . . . are unnecessary: B. Spellberg, "Why You May Not Need All Those Days of Antibiotics," *The Conversation*, July 31, 2017.

28 Llewelyn . . . the length of treatments: Llewelyn, Fitzpatrick, Darwin, et al., "The Antibiotic Course Has Had Its Day."

28 300,000 people suffer from appendicitis: C. J. DeFrances, K. A. Cullen, and L. J. Kozak, "National Hospital Discharge Survey: 2005 Annual Summary with Detailed Diagnosis and Procedure Data," *Vital Health Statistics* 165 (2007): 1–209.

28 In 2008, Italian investigators: A. Basoli, P. Chirletti, E. Cirino, et al., "A Prospective, Double-Blind, Multicenter, Randomized Trial Comparing Ertapenem 3 Vs ≥ 5 Days in Community-Acquired Intraabdominal Infection," *Journal of Gastrointestinal Surgery* 12 (2008): 592–600.

28 In 2015 . . . another five hundred adults: R. G. Sawyer, J. A. Claridge, A. B. Nathens, et al., "Trial of Short-Course Antimicrobial Therapy for Intraabdominal Infection," *New England Journal of Medicine* 372 (2015): 1996–2005.

28 now recommend only four or five days: J. S. Solomkin, J. E. Mazuski, J. S. Bradley, et al., "Diagnosis and Management of Complicated Intra-Abdominal Infections in Adults and Children: Guidelines by the Surgical Infection Society and the Infectious Diseases Society of America," *Clinical Infectious Diseases* 50 (2010): 133–64.

28 Urinary tract infections: A. L. Flores-Mireles, J. N. Walker, M. Caparon, et al., "Urinary Tract Infections: Epidemiology, Mechanisms of

Infection and Treatment Options," *Nature Reviews Microbiology* 13 (2015): 269–84.

28 340 adult women with cystitis: K. Gupta, T. M. Hooten, P. L. Roberts, et al., "Short-Course Nitrofurantoin for the Treatment of Acute Uncomplicated Cystitis in Women," *Archives of Internal Medicine* 167 (2007): 2207–12.

29 In 2018 . . . another 500 women: A. Huttner, A. Kowalczyk, A. Turjeman, et al., "Effect of 5-Day Nitrofurantoin vs. Single-Dose Fosfomycin on Clinical Resolution of Uncomplicated Lower Urinary Tract Infection," *Journal of the American Medical Association* 319 (2018): 1781–89.

29 Advisory groups now recommend: K. Gupta, T. M. Hooton, K. G. Naber, et al., "International Clinical Practice Guidelines for the Treatment of Acute Uncomplicated Cystitis and Pyelonephritis in Women: A 2010 Update by the Infectious Diseases Society of America and the European Society of Microbiology and Infectious Diseases," *Clinical Infectious Diseases* 52 (2011): e103–20.

29 In 2012 . . . women with pyelonephritis: T. Sandberg, G. Skoog, A. B. Hermansson, et al., "Ciprofloxacin for 7 Days Versus 14 Days in Women with Acute Pyelonephritis: A Randomized, Open-Label and Double-Blind, Placebo-Controlled, Non-Inferiority Trial," *Lancet* 380 (2012): 484–90.

29 In 2013 . . . eight trials: N. Eliakim-Raz, D. Yahav, M. Paul, et al., "Duration of Antibiotic Treatment for Acute Pyelonephritis and Septic Urinary Tract Infection—7 Days or Less Versus Longer Treatment: Systematic Review and Meta-Analysis of Randomized Controlled Trials," *Journal of Antimicrobial Chemotherapy* 68 (2013): 2183–291.

29 In 2017 . . . another 270 women: M. T. Fox, M. T. Melia, R. G. Same, et al., "A Seven-Day Course of Trimethoprim-Sulfamethoxazole May Be as Effective as a Seven-Day Course of Ciprofloxacin for the Treatment of Pyelonephritis," *American Journal of Medicine* 130 (2017): 842–45.

29 pyelonephritis can now be treated: Gupta, Hooton, Naber, et al., "International Clinical Practice Guidelines for the Treatment of Acute Uncomplicated Cystitis and Pyelonephritis in Women."

29 In 2009, Greek investigators: M. E. Falagas, D. E. Karageorgopoulos, A. R. Grammatikos, et al., "Effectiveness and Safety of Short vs.

Long Duration of Antibiotic Therapy for Acute Bacterial Sinusitis: A Meta-Analysis of Randomized Trials," *British Journal of Clinical Pharmacology* 67 (2009): 161–71.

29 advisory bodies now recommend: A. W. Chow, M. S. Benninger, I. Brook, et al., "IDSA Clinical Practice Guideline for Acute Bacterial Rhinosinusitis in Children and Adults," *Clinical Infectious Diseases* 54 (2012): e72–112.

29 In 2016, Spanish investigators: A. Uranga, P. P España, A. Bilbao, et al., "Duration of Antibiotic Treatment in Community-Acquired Pneumonia: A Multicenter Randomized Clinical Trial," *JAMA Internal Medicine* 176 (2016): 1257–65.

30 Advisory committees now recommend: L. A. Mandell, R. G. Wunderink, A. Anzueto, et al., "Infectious Diseases Society of America/American Thoracic Society Consensus Guidelines on the Management of Community-Acquired Pneumonia in Adults," *Clinical Infectious Diseases* 44 (2007): S27–72.

30 a hundred thousand cases of cellulitis: G. R. Lum, "Cellulitis Among Active Duty Service Members, US Armed Forces, 1998–2001," *Medical Surveillance Monthly Report* 8 (2001): 6–9.

30 In 2004 . . . treated ninety adults with cellulitis: J. J. Hepburn, D. P. Dooley, P. J. Skidmore, et al., "Comparison of Short-Course (5 Days) and Standard (10 Days) Treatment for Uncomplicated Cellulitis," *Archives of Internal Medicine* 164 (2004): 1669–74.

30 Advisory groups now recommend: D. L. Stevens, A. L. Bisno, H. F. Chambers, et al., "Practice Guidelines for the Diagnosis and Management of Skin and Soft Tissue Infections: 2014 Update by the Infectious Diseases Society of America," *Clinical Infectious Diseases* 59 (2014): 147–59.

30 In 2018 . . . severe bloodstream infections: D. Yahav, E. Franceschini, F. Koppel, et al., "Seven Versus Fourteen Days of Antibiotic Therapy for Uncomplicated Gram-Negative Bacteremia: A Non-Inferiority Randomized Trial," *Clinical Infectious Diseases* (2018), doi: 10.1093/cid/ciy1054.

31 In 2001 . . . 1,300 children with cystitis: D. Tran, D. G. Muchant, and S. C. Aronoff, "Short-Course Versus Conventional Length Antimicrobial Therapy for Uncomplicated Lower Urinary Tract Infections

in Children: A Meta-Analysis of 1279 Patients," *Journal of Pediatrics* 139 (2001): 93–99.

31 in 2002 . . . 650 children with cystitis: M. Michael, E. M. Hodson, J. C. Craig, et al., "Short Compared with Standard Duration of Antibiotic Treatment for Urinary Tract Infection: A Systematic Review of Randomized Controlled Trials," *Archives of Disease in Childhood* 87 (2002): 118–23.

31 Advisory bodies now recommend: J. L. Robinson, J. C. Finlay, M. E. Lang, et al., "Urinary Tract Infection in Infants and Children: Diagnosis and Management," *Paediatrics and Child Health* 19 (2014): 315–25.

31 Ear infections are the single most: S. M. Schappert and E. A. Rechtsteiner, "Ambulatory Medical Care Utilization Estimates for 2007," *Vital Health Statistics* 13 (2011): 1–38.

31 WHO . . . review of existing studies: World Health Organization, "Effectiveness of Shortened Course (≤3 days) of Antibiotics for Treatment of Acute Otitis Media in Children: A Systematic Review of Randomized Controlled Trials," 2009, http://apps.who.int/iris/bitsteam/10665/144177/1/9789241598446_eng.pdf?ua=1&ua=1.

32 something called the microbiome: I. Cho and M. J. Blaser, "The Human Microbiome: At the Interface of Health and Disease," *Nature Reviews Genetics* 13 (2012): 260–70; M. Broadfoot, "A Delicate Balance," *Science* 360 (2018): 18–20.

32 Brad Spellberg . . . Constantine: B. Spellberg, "The New Antibiotic Mantra—'Shorter Is Better,'" *JAMA Internal Medicine* 176 (2016): 1254–55.

33 "In the first five to ten beds": D. E. Rogers, "The Early Years: The Medical World in Which Walsh McDermott Trained," *Daedalus* 115 (1986): 1–18.

Chapter 3: Antibiotic Drops Don't Treat Pinkeye

35 Eyeball licking was: M. Castillo, "Japanese 'Eyeball-Licking' Trend Carries Blindness Risk," CBS News, June 17, 2013.

36 Conjunctivitis is not a trivial infection: A. A. Azari and N. P. Narney, "Conjunctivitis: A Systematic Review of Diagnosis and Treatment," *Journal of the American Medical Association* 310 (2013): 1721–29.

36 Most of these institutions *require*: M. Keen and M. Thompson, "Treatment of Acute Conjunctivitis in the United States and Evidence of Antibiotic Overuse: Isolated Issue or a Systematic Problem?" *Ophthalmology* 124 (2017): 1096–98.

37 Symptoms include: Azari and Narney, "Conjunctivitis."

37 less likely to go to a doctor: Azari and Narney, "Conjunctivitis."

37 conjunctivitis is caused by viruses: Azari and Narney, "Conjunctivitis."

38 simple rule for distinguishing: Azari and Narney, "Conjunctivitis."

38 there's a lot of overlap: Keen and Thompson, "Treatment of Acute Conjunctivitis in the United States and Evidence of Antibiotic Overuse."

38 reviewed all the existing studies: A. Sheikh, B. Hurwitz, C. P. van Schayck, et al., "Antibiotics Versus Placebo for Acute Bacterial Conjunctivitis," *Cochrane Systematic Review,* September 12, 2012, doi: 10.1002/14651858.CD001211.pub3.

39 The study . . . from Oxfordshire: P. W. Rose, A. Harnden, A. B. Brueggemann, et al., "Chloramphenicol Treatment for Acute Infective Conjunctivitis in Children in Primary Care: A Randomised Double-Blind Placebo-Controlled Trial," *Lancet* 366 (2005): 37–43.

39 Switzerland and the Netherlands: R. P. Rietveld, G. ter Riet, P. J. Bindels, et al., "The Treatment of Acute Infectious Conjunctivitis with Fusidic Acid: A Randomised Controlled Trial," *British Journal of General Practice* 55 (2005): 924–30.

39 studies judged by the researchers: I. M. Miller, J. Wittreich, R. Vogel, et al., "The Safety and Efficacy of Topical Norfloxacin Compared with Placebo in the Treatment of Acute, Bacterial Conjunctivitis," *European Journal of Ophthalmology* 2 (1992): 58–66; B. E. Silverstein, C. Allaire, K. M. Bateman, et al., "Efficacy and Tolerability of Besifloxacin Ophthalmic Suspension 0.6% Administered Twice Daily for 3 Days in the Treatment of Bacterial Conjunctivitis: A Multicenter, Randomized, Double-Masked, Vehicle-Controlled, Parallel-Group Study in Adults and Children," *Clinical Therapeutics* 33 (2011): 13–26; M. E. Tepedino, W. H. Heller, D. W. Usher, et al., "Phase III Efficacy and Safety Study of Besifloxacin Ophthalmic Suspension 0.6% in the Treatment of Bacterial Conjunctivitis," *Current Medical Research and*

Opinion 25 (2009): 1158–69; F. Gigliotti, J. O. Hendley, J. Morgan, et al., "Efficacy of Topical Antibiotic Therapy in Acute Conjunctivitis in Children," *Journal of Pediatrics* 104 (1984): 623–26.

40 American Academy of Ophthalmology now recommends: L. Meszaros, "Antibiotic Drops for Pink Eye Are Seldom Necessary, but Widely Prescribed," MDLinx, June 30, 2017, https://www.mdlinx.com/ophthalmology/article/932.

40 about 60 percent of people in the United States: N. S. Shekhawat, R. M. Shtein, T. S. Blachley, et al., "Antibiotic Prescription Fills for Acute Conjunctivitis in a Large United States Managed Care Network," *Ophthalmology* 124 (2017): 1099–107.

40 Antibiotic eye drops can be sensitizing: Choosing Wisely, "Antibiotics for Pink Eye: When You Need Them—and When You Don't," http://www.choosingwisely.org/patient-resources/antibiotics-for-pink-eye.

40 can promote the emergence of resistant strains: Azari and Narney, "Conjunctivitis."

40 antibiotic-steroid combination: Shekhawat, Shtein, Blachley, et al., "Antibiotic Prescription Fills."

40 Steroids . . . will only prolong: Shekhawat, Shtein, Blachley, et al., "Antibiotic Prescription Fills."

40 Antibiotics change the types and quantities of bacteria: L. J. Lu and J. Liu, "Human Microbiota and Ophthalmic Disease," *Yale Journal of Biology and Medicine* 89 (2016): 325–30.

40 Antibiotic eye drops aren't free: Choosing Wisely, "Antibiotics for Pink Eye."

41 eyeball licking wasn't really a thing: D. Mikkelson, "Japanese Eyeball-Licking Craze," Snopes, August 8, 2013.

Chapter 4: Vitamin D Supplements Aren't a Cure-all

45 vitamin D craze: Gina Kolata, "Why Are So Many People Popping Vitamin D?" *New York Times,* April 10, 2017; J. McDougall, "Vitamin D Supplements Are Harmful—Sunshine and Food Determine Health," *Forks Over Knives,* May 14, 2015, https://www.forksoverknives.com/vitamin-d-supplements-are-harmful-sunshine-and-food-determine-health/#gs.Sq8KFrY.

45 Vitamin D now outsells every other supplement: C. Price, "Our Bodies Make Vitamin D Naturally: So Why Do Americans Spend Billions on Supplements?" *Los Angeles Times*, May 16, 2016.

46 Dr. Clifford Rosen . . . "a religion": Kolata, "Why Are So Many People Popping Vitamin D?"

46 Vitamin D facilitates the absorption of calcium: M. Holick, "Vitamin D Deficiency," *New England Journal of Medicine* 357 (2007): 266–81.

47 Rickets . . . has essentially vanished: L. R. Johnson, "Vitamin D Insufficiency Due to Insufficient Exposure to Sunlight and Related Pathology," *Inquiries* (2010), http://www.inquiriesjournal.com/articles/346/vitamin-d-insufficiency-due-to-insufficient-exposure-to-sunlight-and-related-pathology.

47 Humankind's debut . . . obtained only from food: Johnson, "Vitamin D Insufficiency."

48 more likely to develop colon, breast, and prostate cancer: Johnson, "Vitamin D Insufficiency."

48 study of thirty thousand people: P. G. Lindqvist, E. Epstein, K. Nielsen, et al., "Avoidance of Sun Exposure as a Risk Factor for Major Causes of Death: A Competing Risk Analysis of the Melanoma in Southern Sweden Cohort," *Journal of Internal Medicine* 280 (2016): 375–87.

48 Observational studies: S. B. Mohr, E. D. Gorham, J. Kim, et al., "Meta-Analysis of Vitamin D Sufficiency for Improving Survival of Patients with Breast Cancer," *Anticancer Research* 34 (2014): 1163–66; Y. Dror, S. M. Giveon, M. Hoshen, et al., "Vitamin D Levels for Preventing Acute Coronary Syndrome and Mortality: Evidence of a Non-Linear Association," *Journal of Clinical Endocrinology and Metabolism* 98 (2013): 2160–67; J. Kumar, P. Munter, F. J. Kaskel, et al., "Prevalence and Associations of 25-Hydroxyvitamin D Deficiency in US Children: NHANES 2001–2004," *Pediatrics* (2009), http://pediatrics.aappublications.org/content/124/3/e362; J. H. Jhee, H. Kim, S. Park, et al., "Vitamin D Deficiency Is Significantly Associated with Depression in Patients with Chronic Kidney Disease," *PLOS One* 12 (2017), doi: 10.1371/journal.pone.0171009; M. Holick, "Vitamin D and Sunlight: Strategies for Cancer Prevention and Other Health Benefits," *Clinical Journal of the American Society for*

Nephrology 3 (2008): 1548–54; R. P. Heaney, "Vitamin D in Health and Disease," *Clinical Journal of the American Society of Nephrology* 3 (2008): 1535–41; T. Kulie, A. Groff, J. Redmer, et al., "Vitamin D: An Evidence-Based Review," *Journal of the American Board of Family Medicine* 22 (2009): 698–706; R. Nair and A. Maseeh, "Vitamin D: The 'Sunshine' Vitamin," *Journal of Pharmacology and Pharmacotherapeutics* 3 (2012): 118–26; P. S. Nijjar, "Vitamin D and Cardiovascular Disease: Where We Currently Are," June 8, 2015, American College of Cardiology, http://www.acc.org/latest-in-cardiology/articles/2015/06/08/12/06/vitamin-d-and-cardiovascular-disease-where-we-currently-are.

49 vitamin D binds to cells: C. Kresser, "Vitamin D: More Is Not Better," August 2, 2016, https://chriskresser.com/vitamin-d-more-is-not-better/; Holick, "Vitamin D and Sunlight"; Holick, "Vitamin D Deficiency"; correspondence, *New England Journal of Medicine* 357 (2007): 1980–82; A. S. Dusso, A. J. Brown, and E. Slatopolsky, "Vitamin D," *American Journal of Physiology Renal Physiology* 289 (2005): F8–F28.

49 In 2019, researchers from Tokyo: M. Urashima, H. Ohdaira, T. Akutsu, et al., "Effect of Vitamin D Supplementation on Relapse-Free Survival Among Patients with Digestive Tract Cancers: The AMATERASU Randomized Clinical Trial," *Journal of the American Medical Association* 321 (2019): 1361–69.

50 In 2014, researchers from Denmark: G. Bjelakovic, L. L. Gluud, D. Nikolova, et al., "Vitamin D Supplementation for Prevention of Cancer in Adults," *Cochrane Data Base System Reviews* (2014), doi: 10.1002/14651858.CD007469.pub2.

50 In 2017, researchers from Creighton: J. Lappe, P. Watson, D. Travers-Gustafson, et al., "Effect of Vitamin D and Calcium Supplementation on Cancer Incidence in Older Women: A Randomized Clinical Trial," *Journal of the American Medical Association* 317 (2017): 1234–43.

50 In 2018, researchers from the United Kingdom: B. Goulão, F. Stewart, J. A. Ford, et al., "Cancer and Vitamin D Supplementation: A Systematic Review and Meta-Analysis," *American Journal of Clinical Nutrition* 107 (2018): 652–63.

50 In 2019, researchers from Harvard Medical School: J. E. Manson, N. R. Cook, I.-M. Lee, et al., "Vitamin D Supplements and Prevention

of Cancer and Cardiovascular Disease," *New England Journal of Medicine* 380 (2019): 33–44.

50 In 2014, researchers from New Zealand: M. J. Bolland, A. Grey, G. D. Gamble, et al., "The Effect of Vitamin D Supplementation on Skeletal, Vascular, or Cancer Outcomes: A Trial Sequential Meta-Analysis," *Lancet Diabetes and Endocrinology* 2 (2014): 307–20.

51 In 2012, researchers from New Zealand: D. R. Murdoch, S. Slow, S. T. Chambers, et al., "Effect of Vitamin D_3 Supplementation on Upper Respiratory Tract Infections in Healthy Adults: The VIDARIS Randomized Controlled Trial," *Journal of the American Medical Association* 308 (2012): 1333–39.

51 In 2017, French researchers reviewed: P. Autier, P. Mullie, A. Macacu, et al., "Effect of Vitamin D Supplementation on Non-Skeletal Disorders: A Systematic Review of Meta-Analyses and Randomized Trials," *Lancet Diabetes and Endocrinology* 5 (2017): 986–1004.

51 In 2010, Australian researchers: K. M. Sanders, A. L. Stuart, E. J. Williamson, et al., "Annual High-Dose Oral Vitamin D and Falls and Fractures in Older Women: A Randomized Controlled Trial," *Journal of the American Medical Association* 303 (2010): 1815–22.

51 In 2014, French researchers: Bolland, Grey, Gamble, et al., "The Effect of Vitamin D Supplementation on Skeletal, Vascular, or Cancer Outcomes."

51 In 2017, researchers in China: J.-G. Zhao, X.-T. Zeng, J. Wang, et al., "Association Between Calcium or Vitamin D Supplementation and Fracture Incidence in Community-Dwelling Older Adults: A Systematic Review and Meta-Analysis," *Journal of the American Medical Association* 318 (2017): 2466–82.

52 In 2018, researchers from New Zealand: M. J. Bolland, A. Grey, and A. Avenell, "Effects of Vitamin D Supplementation on Musculoskeletal Health: A Systematic Review, Meta-Analysis, and Trial Sequential Analysis," *Lancet Diabetes and Endocrinology* 6 (2018): 847–58.

52 vitamin D could cause: Kolata, "Why Are So Many People Popping Vitamin D?"

52 In observational studies, individuals decide: J. Belluz, "Your Vitamin D Tests and Supplements Are Probably a Waste of Money," Vox, June 20, 2017.

52 Vitamin D is a marker for poor health: P. Autier, M. Boniol, C. Pizot, et al., "Vitamin D Status and Ill Health: A Systematic Review," *Lancet Diabetes and Endocrinology* 2 (2014): 76–89.

53 vitamin D levels are consistently lower in black people: D. E. Powe, M. K. Evans, J. Wenger, et al., "Vitamin D–Binding Protein and Vitamin D Status of Black Americans and White Americans," *New England Journal of Medicine* 369 (2013): 1991–2000.

53 The failure of physicians . . . at the same time: Holick, "Vitamin D Deficiency."

54 sunlight also enhances the production of a variety of other substances: Kresser, "Vitamin D: More Is Not Better."

54 two federal advisory bodies: J. Corliss, "Vitamin D Testing Not Recommended for Most People," *Harvard Health Blog*, November 26, 2014, https://www.health.harvard.edu/blog/vitamin-d-testing-recommended-people-201411267547; Belluz, "Your Vitamin D Tests and Supplements Are Probably a Waste of Money"; J. E. Manson and S. S. Bassuk, "Vitamin D Research and Clinical Practice: At a Crossroads," *Journal of the American Medical Association* 313 (2015): 1311–12; Kresser, "Vitamin D: More Is Not Better"; Nair and Maseeh, "Vitamin D: The 'Sunshine' Vitamin."

55 Gina Kolata pulled back the curtain: Kolata, "Why Are So Many People Popping Vitamin D?"

56 embraced by Gwyneth Paltrow's lifestyle and wellness company, Goop: L. Szabo, "Vitamin D, the Sunshine Supplement, Has Shadowy Money Behind It," *New York Times*, August 18, 2018.

56 "Demand for vitamin D testing": Szabo, "Vitamin D, the Sunshine Supplement."

56 Liz Szabo, in an article for the *New York Times*: Szabo, "Vitamin D, the Sunshine Supplement."

56 Armstrong, in an article for the *Wall Street Journal*: D. Armstrong, "Researcher Received Industry Funds," *Wall Street Journal*, April 18, 2008.

57 Holick was asked to step down: Szabo, "Vitamin D, the Sunshine Supplement."

57 Holick argued that his industry funding: L. Szabo, "The Man Who Sold America on Vitamin D—and Profited in the Process," *Kaiser Health News*, August 20, 2018.

57 Advocacy organizations . . . promote the notion: A. Grey, "Web of Industry, Advocacy, and Academia in the Management of Osteoporosis," *British Medical Journal* (2015), doi: 10.1136/bmj.h3170.

58 "A lot of clinicians . . . screen everyone": Kolata, "Why Are So Many People Popping Vitamin D?"

Chapter 5: Supplemental Antioxidants Increase the Risk of Cancer and Heart Disease

59 Oxidation . . . antioxidants: National Cancer Institute, "Antioxidants and Cancer Prevention," https://www.cancer.gov/causes-prevention/risk/diet/antioxidants-fact-sheet#q3.

60 researchers from Oslo, Norway: E. Bjelke, "Dietary Vitamin A and Human Lung Cancer," *International Journal of Cancer* 15 (1975): 561–65.

60 Two subsequent studies: R. B. Shekelle, S. Liu, W. J. Raynor Jr., et al., "Dietary Vitamin A and Risk of Cancer in the Western Electric Study," *Lancet* 2 (1981): 1185–89; G. Kvale, E. Bjelke, and J. J. Gart, "Dietary Habits and Lung Cancer Risk," *International Journal of Cancer* 31 (1983): 397–405.

60 National Cancer Institute: R. G. Ziegler, "A Review of Epidemiologic Evidence That Carotenoids Reduce Risk of Cancer," *Journal of Nutrition* 119 (1989): 116–22.

60 more than two hundred human studies: J. H. Weisburger, "Nutritional Approaches to Cancer Prevention with Emphasis on Vitamins, Antioxidants, and Carotenoids," *American Journal of Clinical Nutrition* 53 (1991): 226S–37S; K. A. Steinmetz and J. D. Potter, "Vegetables, Fruit, and Cancer. I. Epidemiology," *Cancer Causes and Control* 2 (1991): 325–57; G. Block, B. Patterson, and A. Subar, "Fruit, Vegetables, and Cancer Prevention: A Review of Epidemiological Evidence," *Nutrition and Cancer* 18 (1992): 1–29.

60 researchers from the World Cancer Research Fund: K. A. Steinmetz and J. D. Potter, "Vegetables, Fruit, and Cancer Prevention: A Review," *Journal of the American Dietetic Association* 96 (1996): 1027–39.

61 researchers from Harvard Medical School: C. H. Hennekens, J. E. Buring, J. E. Manson, et al., "Lack of Effect of Long-Term Supplementation with Beta Carotene on the Incidence of Malignant Neoplasms and Cardiovascular Disease," *New England Journal of Medicine* 334 (1996): 1145–49.

61 researchers from Helsinki, Finland: Alpha-Tocopherol Beta Carotene Cancer Prevention Study Group, "The Effect of Vitamin E and Beta Carotene on the Incidence of Lung Cancer and Other Cancers in Male Smokers," *New England Journal of Medicine* 330 (1994): 1029–35; M. T. Rautalahti, J. R. K. Virtamo, P. R. Taylor, et al., "The Effects of Supplementation with Alpha-Tocopherol and Beta-Carotene on the Incidence and Mortality of Carcinoma of the Pancreas in a Randomized, Controlled Trial," *Cancer* 86 (1999): 37–42; J. Virtamo, B. K. Edwards, M. Virtanen, et al., "Effects of Supplemental Alpha-Tocopherol and Beta-Carotene on Urinary Tract Cancer: Incidence and Mortality in a Controlled Trial (Finland)," *Cancer Causes and Control* 11 (2000): 933–39; E. Albanes, M. Malila, P. R. Taylor, et al., "Effects of Supplemental Alpha-Tocopherol and Beta-Carotene on Colorectal Cancer: Results from a Controlled Trial (Finland)," *Cancer Causes and Control* 11 (2000): 197–205; M. E. Wright, J. Virtamo, A. M. Hartman, et al., "Effects of Alpha-Tocopherol and Beta-Carotene Supplementation on Upper Aerodigestive Tract Cancers in a Large, Randomized Controlled Trial," *Cancer* 109 (2007): 891–98.

61 researchers from Brigham and Women's Hospital: I.-M. Lee, N. R. Cook, J. E. Manson, et al., "Beta-Carotene Supplementation and Incidence of Cancer and Cardiovascular Disease: The Women's Health Study," *Journal of the National Cancer Institute* 91 (1999): 2102–6.

62 researchers found that vitamin E: I.-M. Lee, N. R. Cook, J. M. Gaziano, et al., "Vitamin E in the Primary Prevention of Cardiovascular Disease and Cancer: The Women's Health Study: A Randomized Controlled Trial," *Journal of the American Medical Association* 294 (2005): 56–65.

62 researchers from Ontario, Canada: Heart Outcomes Prevention Evaluation Study Investigators, "Vitamin E Supplementation and Cardiovascular Events in High-Risk Patients," *New England Journal of Medicine* 342 (2000): 154–60; HOPE and HOPE-TOO Trial Investigators, "Effects of Long-Term Vitamin E Supplementation on Cardiovascular Events and Cancer," *Journal of the American Medical Association* 293 (2005): 1338–47.

62 researchers from Copenhagen: G. Bjelakovic, D. Nikolova, R. G. Simonetti, et al., "Antioxidant Supplements for Prevention of Gastrointestinal Cancers: A Systematic Review and Meta-Analysis," *Lancet* 364 (2004): 1219–28.

62 researchers in Paris: S. Hercberg, P. Galan, P. Preziosi, et al., "The SU.VI.MAX Study: A Randomized, Placebo-Controlled Trial of the Health Effects of Antioxidant Vitamins and Minerals," *Archives of Internal Medicine* 164 (2004): 2335–42.

62 researchers from Brigham and Women's Hospital and Harvard Medical School: J. M. Gaziano, R. J. Glynn, W. G. Christen, et al., "Vitamin E and C in the Prevention of Prostate and Total Cancer in Men: The Physicians' Health Study II Randomized Controlled Trial," *Journal of the American Medical Association* 301 (2009): 52–62.

63 U.S. Preventive Services Task Force: S. P. Fortmann, B. U. Burda, C. A. Senger, et al., "Vitamin and Mineral Supplements in the Primary Prevention of Cardiovascular Disease and Cancer: An Updated Systematic Evidence Review for the U.S. Preventive Services Task Force," *Annals of Internal Medicine* 159 (2013): 824–34.

63 "A protective effect of diet": B. Halliwell, "The Antioxidant Paradox," *Lancet* 355 (2000): 1179–80.

63 these foods contain other phytochemicals: National Cancer Institute, "Antioxidants and Cancer Prevention," https://www.cancer.gov/causes-prevention/risk/diet/antioxidants-fact-sheet#q3.

64 researchers from the Fred Hutchinson Cancer Research Center: G. S. Omenn, G. E. Goodman, M. D. Thornquist, et al., "Effects of a Combination of Beta Carotene and Vitamin A on Lung Cancer and Cardiovascular Disease," *New England Journal of Medicine* 334 (1996): 1150–55.

64 Six years after: G. E. Goodman, M. D. Thornquist, J. Balmes, et al., "The Beta-Carotene and Retinol Efficacy Trial: Incidence of Lung Cancer and Cardiovascular Disease Mortality During 6-Year Follow-Up After Stopping Beta-Carotene and Retinol Supplements," *Journal of the National Cancer Institute* 96 (2004): 1743–50.

64 Eleven years later: M. L. Neuhouser, M. J. Barnett, A. R. Kristal, et al., "Dietary Supplement Use and Prostate Cancer Risk in the Carotene and Retinol Efficacy Trial," *Cancer Epidemiology Biomarkers and Prevention* 18 (2009): 2202–6.

65 Johns Hopkins School of Medicine . . . reviewed nineteen trials: E. R. Miller, R. Pastor-Barriuso, D. Dalai, et al., "Meta-Analysis: High-Dosage Vitamin E Supplementation May Increase All-Cause Mortality," *Annals of Internal Medicine* 142 (2005): 37–46.

65 researchers from Laval University Cancer Research Center: F. Meyer, P. Galan, P. Douville, et al., "Antioxidant Vitamin and Mineral Supplementation and Prostate Cancer Prevention in the SU.VI.MAX Trial," *International Journal of Cancer* 116 (2005): 182–86.

65 In 2007, French researchers: S. Hercberg, K. Ezzedine, C. Guinot, et al., "Antioxidant Supplementation Increases the Risk of Skin Cancers in Women but Not in Men," *Journal of Nutrition* 137 (2007): 2098–105.

66 "It clearly appears more cautious": S. Hercberg, E. Kesse-Guyot, N. Druesne-Pecollo, et al., "Incidence of Cancers, Ischemic Cardiovascular Diseases and Mortality During 5-Year Follow-Up After Stopping Antioxidant Vitamins and Minerals Supplements: A Postintervention Follow-Up in the SU.VI.MAX Study," *International Journal of Cancer* 127 (2010): 1875–81.

66 researchers from the Cleveland Clinic: E. A. Klein, I. M. Thompson, C. M. Tangen, et al., "Vitamin E and the Risk of Prostate Cancer: The Selenium and Vitamin E Cancer Prevention Trial (SELECT)," *Journal of the American Medical Association* 306 (2011): 1549–56.

66 the Centers for Disease Control: M. L. Watkins, J. D. Erickson, M. J. Thun, et al., "Multivitamin Use and Mortality in a Large Prospective Study," *American Journal of Epidemiology* 152 (2000): 149–62.

66 performed by the American Cancer Society: V. L. Stevens, M. L. McCullough, W. R. Diver, et al., "Use of Multivitamins and Prostate Cancer Mortality in a Large Cohort of US Men," *Cancer Causes and Control* 16 (2005): 643–50.

67 National Cancer Institute: K. A. Lawson, M. E. Wright, A. Subar, et al., "Multivitamin Use and Risk of Prostate Cancer in the National Institutes of Health–AARP Diet and Health Study," *Journal of the National Cancer Institute* 99 (2007): 754–64.

67 Studies of experimental animals: E. Piskounova, M. Agathocleous, M. M. Murphy, et al., "Oxidative Stress Inhibits Distant Metastasis by Human Melanoma Cells," *Nature* 527 (2015): 186–91; K. Le Gal, M. Ibrahim, E. Larsson, et al., "Antioxidants Can Increase Melano-

mas in Mice," *Science Translational Medicine* 7, no. 308 (2015): 308re; V. I. Sayin, M. X. Ibrahim, E. Larsson, et al., "Antioxidants Accelerate Lung Cancer Progression in Mice," *Science Translational Medicine* 6, no. 221 (2014): 221ra15.

69 "The thing to do with [these studies] is just ride them out": S. S. Wang, "Is This the End of Popping Vitamins?" *Wall Street Journal*, October 25, 2011.

69 "intravenous lounges": M. Thielking, "Vitamin IVs Promise to Erase Jet Lag and Clear Your Mind. Where's the Evidence?" *STAT*, February 23, 2017.

Chapter 6: Testosterone for "Low T" Is Dangerous and Unnecessary

71 Symptoms of hypogonadism: P. Kumar, N. Kumar, D. S. Thaku, et al., "Male Hypogonadism: Symptoms and Treatment," *Journal of Advanced Pharmaceutical Technology and Research* 1 (2010): 297–301.

73 "All-Star Large Pharma Marketing Team of the Year": L. Dobrow, "2013 All-Star Large Pharma Marketing Team of the Year: Androgel," *Medical Marketing and Media*, January 2, 2013.

74 Schwartz and Woloshin: N. Singer, "Selling That New-Man Feeling," *New York Times*, November 23, 2013; A. O'Connor, "Men's Use of Testosterone on the Rise," *New York Times*, June 3, 2013; S. Perry, "'Low T' Campaign and Other Disease Mongering Can Do Harm, Doctors Say," *MinnPost*, June 4, 2013; R. Rettner, "Testosterone Prescriptions Nearly Triple in Last Decade," *Live Science*, June 3, 2013; E. H. Gan, S. Pattman, S. H. S. Pearce, et al., "A UK Epidemic of Testosterone Prescribing, 2001–2010," *Clinical Endocrinology* 79 (2013): 564–70.

74 In 1997, researchers from St. Louis University Health Sciences Center: R. Sih, J. E. Morley, F. E. Kaiser, et al., "Testosterone Replacement in Older Hypogonadal Men: A 12-Month Randomized Controlled Trial," *Journal of Clinical Endocrinology and Metabolism* 82 (1997): 1661–67.

74 In 2000, researchers at the University of Trier: O. T. Wolf, R. Preut, D. H. Hellmamer, et al., "Testosterone and Cognition in Elderly Men: A Single Testosterone Injection Blocks the Practice Effect in Verbal Fluency, but Has No Effect on Spatial or Verbal Memory," *Biological Psychiatry* 1 (2000): 650–54.

74 In 2006, researchers from the David Geffen School of Medicine: P. H. Lu, D. A. Masterman, R. Mulnard, et al., "Effects of Testosterone on Cognition and Mood in Male Patients with Mild Alzheimer Disease and Healthy Elderly Men," *Archives of Neurology* 63 (2006): 177–85.

75 In 2000, researchers at Hartford Hospital: P. Reddy, C. M. White, A. B. Dunn, et al., "The Effect of Testosterone on Health-Related Quality of Life in Elderly Males—A Pilot Study," *Journal of Pharmaceutics and Therapeutics* 25 (2000): 421–26.

75 In 2005, researchers from the University of Adelaide: M. T. Haren, G. A. Wittert, I. M. Chapman, et al., "Effect of Oral Testosterone Undecanoate on Visuospatial Cognition, Mood and Quality of Life in Elderly Men with Low-Normal Gonadal Status," *Maturitas* (2005) 14: 124–33.

75 In 2001, researchers from Columbia University: S. N. Seidman, E. Spatz, C. Rizzo, et al., "Testosterone Replacement Therapy for Hypogonadal Men with Major Depressive Disorder: A Randomized, Placebo-Controlled Clinical Trial," *Journal of Clinical Psychiatry* 62 (2001): 406–12.

75 In 2010, researchers from McLean Hospital: H. G. Pope Jr., R. Amiaz, B. P. Brennan, et al., "Parallel-Group Placebo-Controlled Trial of Testosterone Gel in Men with Major Depressive Disorder Displaying an Incomplete Response to Standard Antidepressant Treatment," *Journal of Clinical Psychopharmacology* 30 (2010): 126–34.

76 In 2006, researchers from the University of Florida: M. S. Okun, H. H. Fernandez, R. L. Rodriguez, et al., "Testosterone Therapy in Men with Parkinson Disease: Results of the TEST-PD Study," *Archives of Neurology* 63 (2006): 729–35.

76 In 2008, researchers from the University Medical Center in Utrecht: M. H. Emmelot-Vonk, H. J. Verhaar, H. R. Nakhai Pour, et al., "Effect of Testosterone Supplementation on Functional Mobility, Cognition, and Other Parameters in Older Men: A Randomized Controlled Trial," *Journal of the American Medical Association* 299 (2008): 39–52.

76 In 1979, researchers at the Center for Maternal and Child Health: O. Benkert, W. Witt, W. Adam, et al., "Effects of Testosterone Undecanoate on Sexual Potency and the Hypothalamic-Pituitary-Gonadal Axis of Impotent Males," *Archives of Sexual Behavior* 8 (1979): 471–79.

77 In 1997, researchers from Mount Sinai School of Medicine: R. C. Schiavi, D. White, J. Mandell, et al., "Effect of Testosterone Administration on Sexual Behavior and Mood in Men with Erectile Dysfunction," *Archives of Sexual Behavior* 26 (1997): 231–41.

77 In 2012, researchers at Boston University School of Medicine: M. Spitzer, S. Basaria, T. G. Travison, et al., "Effect of Testosterone Replacement on Response to Sildenafil Citrate in Men with Erectile Dysfunction: A Parallel, Randomized Trial," *Annals of Internal Medicine* 157 (2012): 681–91.

77 In 2014 . . . researchers from Heidelberg, Australia: E. J. Gianatti, P. Dupuis, R. Hoermann, et al., "Effect of Testosterone Treatment on Constitutional and Sexual Symptoms in Men with Type 2 Diabetes in a Randomized, Placebo-Controlled Clinical Trial," *Journal of Clinical Endocrinology and Metabolism* 99 (2014): 3821–28.

77 156 studies of testosterone: S. Huo, A. R. Scialli, S. McGarvey, et al., "Treatment of Men for 'Low Testosterone': A Systematic Review," *PLOS One* (2018), doi: 10.1371/journal.pone.0162480.

78 testosterone was known to increase anger, aggression, and hostility: E. M. Kouri, S. E. Lukas, H. G. Pope Jr., et al., "Increased Aggressive Responding in Male Volunteers Following the Administration of Gradually Increasing Doses of Testosterone Cypionate," *Drug Alcohol Dependence* 40 (1995): 73–79; R. C. Daly, T. P. Su, P. J. Schmidt, et al., Cerebrospinal Fluid and Behavioral Changes After Methyltestosterone Administration: Preliminary Findings," *Archives of General Psychiatry* 58 (2001): 172–77; J. M. Dabbs, A. E. Karpas, N. Kyomina, et al., "Experimental Raising and Lowering of Testosterone Level Affects Mood in Normal Men and Women," *Social Behavior and Personality* 30 (2002): 795–806; D. B. O'Connor, J. Archer, and F. C. Wu, "Effects of Testosterone on Mood, Aggression, and Sexual Behavior in Young Men: A Double-Blind, Placebo-Controlled, Cross-Over Study," *Journal of Clinical Endocrinology and Metabolism* 89 (2004): 2837–45; H. G. Pope Jr., R. Amiaz, B. P. Brennan, et al., "Parallel-Group Placebo-Controlled Trial of Testosterone Gel in Men with Major Depressive Disorder Displaying an Incomplete Response to Standard Antidepressant Treatment," *Journal of Clinical Psychopharmacology* 30 (2010): 126–34.

78 The poster boy for "roid rage" was Lyle Alzado: D. Anderson, "'Certain Steroid' Reduces Rage of Alzado to Dying Whimper," *Baltimore Eve-*

ning Sun, July 2, 1991; M. Puma, "Roid Rage," ESPN.com, December 23, 2003.

79 In 2010, researchers from Boston University School of Medicine: S. Basaria, A. D. Coviello, T. G. Travison, et al., "Adverse Events Associated with Testosterone Administration," *New England Journal of Medicine* 36 (2010): 109–22.

79 investigators from the University of Texas Southwestern Medical Center: R. Vigen, C. I. O'Donnell, A. E. Barón, et al., "Association of Testosterone Therapy with Mortality, Myocardial Infarction, and Stroke in Men with Low Testosterone Levels," *Journal of the American Medical Association* 310 (2013): 1829–36.

80 researchers from the University of Hong Kong: L. Xu, G. Freeman, B. J. Cowling, et al., "Testosterone Therapy and Cardiovascular Events Among Men: A Systematic Review and Meta-Analysis of Placebo-Controlled Randomized Trials," *BMC Medicine* (2013), http://www.biomedcentral.com/1741-7015/11/108.

80 in 2014, researchers from the University of California, Los Angeles, and the National Institutes of Health: W. D. Finkle, S. Greenland, G. K. Ridgeway, et al., "Increased Risk of Non-Fatal Myocardial Infarction Following Testosterone Therapy Prescription in Men," *PLOS One* (2014), https://journals.plos.org/plosone/article?id=10.1371.pone.0085805.

80 Then the FDA stepped in: Food and Drug Safety Communication, "FDA Cautions About Using Testosterone Products for Low Testosterone Due to Aging; Requires Labeling Change to Inform of Possible Increased Risk of Heart Attack and Stroke with Use," https://www.fda.gov/Drugs/DrugSafety/ucm436259.htm, accessed on January 8, 2019.

81 Endocrine Society . . . sent a letter: Endocrine Society, "The Risk of Cardiovascular Events in Men Receiving Testosterone Therapy: An Endocrine Society Statement," February 7, 2014; Endocrine Society, "Too Many Men Take Testosterone When They Don't Need It," *Washington Post*, January 13, 2014.

82 estrogen . . . sales exceeded $1 billion: M. D. Landau, "The Wild History of Women's Hormone Therapy," *Everyday Health*, August 20, 2018.

82 Women's Health Initiative: J. E. Rossouw, G. L. Anderson, R. L. Prentice, et al., "Risks and Benefits of Estrogen Plus Progestin in Healthy Postmenopausal Women: Principal Results from the Women's Health Initiative Randomized Controlled Trial," *Journal of the American Medical Association* 288 (2002): 321–33; G. L. Anderson, M. Limacher, A. R. Assaf, et al., "Effects of Conjugated Equine Estrogen in Postmenopausal Women with Hysterectomy: The Women's Health Initiative Randomized Controlled Trial," *Journal of the American Medical Association* 291 (2004): 1701–12.

82 increased risk of uterine cancer: D. Grady, T. Bebretasadik, K. Kerlikowski, et al., "Hormone Replacement Therapy and Endometrial Cancer Risk: A Meta-Analysis," *Obstetrics and Gynecology* 85 (1995): 304–13.

Chapter 7: Baby Aspirin Doesn't Prevent First Strokes or First Heart Attacks

86 the drug reinvented itself: M. J. R. Desborough and D. M. Keeling, "The Aspirin Story—From Willow to Wonder Drug," *British Journal of Haematology* 177 (2017): 674–83.

87 In 1974, researchers from Cardiff University: P. C. Elwood, A. L. Cochrane, M. L. Burr, et al., "A Randomized Controlled Trial of Acetyl Salicylic Acid in the Secondary Prevention of Mortality from Myocardial Infarction," *British Medical Journal* 1 (1974): 436–40.

87 In 1987, researchers from Radcliffe University: C. Baigent, R. Collins, P. Appleby, et al., "ISIS-2: 10 Year Survival Among Patients with Suspected Acute Myocardial Infarction in Randomised Comparison of Intravenous Streptokinase, Oral Aspirin, Both, or Neither," *British Medical Journal* 316 (1998): 1337–43.

87 In 2002, these same researchers reviewed 287 studies: Antithrombotic Trialists' Collaboration, "Collaborative Meta-Analysis of Randomised Trials of Antiplatelet Therapy for Prevention of Deaths, Myocardial Infarction, and Stroke in High Risk Patients," *British Medical Journal* 324 (2002): 71–86.

87 In 2009, these same researchers reviewed sixteen more studies: Antithrombotic Trialists' Collaboration, "Aspirin in the Primary and Secondary Prevention of Vascular Disease: Collaborative Meta-Analysis of Individual Participant Data from Randomised Trials," *Lancet* 373 (2009): 1849–60.

88 In 1989, researchers from Harvard Medical School: Steering Committee of the Physicians' Health Study Research Group, "Final Report on the Aspirin Component of the Ongoing Physicians' Health Study," *New England Journal of Medicine* 321 (1989): 129–35.

88 In 2008, researchers from Kumamoto University: H. Ogawa, M. Nakayama, T. Morimoto, et al., "Low-Dose Aspirin for Primary Prevention of Atherosclerotic Events in Patients with Type 2 Diabetes," *Journal of the American Medical Association* 300 (2008): 2134–41.

88 In 2010, researchers from the University of Glasgow: F. G. R. Fowkes, J. F. Price, M. C. W. Stewart, et al., "Aspirin for Prevention of Cardiovascular Events in a General Population Screened for a Low Ankle Brachial Index: A Randomized Controlled Trial," *Journal of the American Medical Association* 303 (2010): 841–48.

88 In 2014, researchers from Waseda University: Y. Ikeda, K. Shimada, T. Teramoto, et al., "Low-Dose Aspirin for Primary Prevention of Cardiovascular Events in Japanese Patients 60 Years or Older with Atherosclerotic Risk Factors," *Journal of the American Medical Association* 312 (2014): 2510–20.

89 In 2016, investigators from Oxford, England: ASCEND Study Collaborative Group, "Effects of Aspirin for Primary Prevention in Persons with Diabetes Mellitus," *New England Journal of Medicine* 379 (2018): 1529–39.

89 In 2018, researchers from Monash University: J. J. McNeil, M. R. Nelson, R. L. Woods, et al., "Effect of Aspirin on All-Cause Mortality in the Healthy Elderly," *New England Journal of Medicine* 379 (2018): 1519–28; J. J. McNeil, R. Wolfe, R. L. Woods, et al., "Effect of Aspirin on Cardiovascular Events and Bleeding in the Healthy Elderly," *New England Journal of Medicine* 379 (2018): 1509–18; J. J. McNeil, R. L. Woods, M. R. Nelson, et al., "Effect of Aspirin on Disability-Free Survival in the Healthy Elderly," *New England Journal of Medicine* 379 (2018): 1499–508.

89 Faculty of Medicine at Imperial College London reviewed thirteen more studies: S. L. Zheng and A. J. Roddick, "Association of Aspirin Use for Primary Prevention with Cardiovascular Events and Bleeding Events: A Systematic Review and Meta-analysis," *Journal of the American Medical Association* 321 (2019): 277–87.

90 much has changed during the past thirty years: P. M. Ridker, "Should Aspirin Be Used for Primary Prevention in the Post-Statin Era?" *New England Journal of Medicine* 379 (2018): 1572–74.

90 In 2010, researchers from the University of Oxford: P. A. Rothwell, M. Wilson, C. E. Elwin, et al., "Long-Term Effect of Aspirin on Colorectal Cancer Incidence and Mortality: 20-Year Follow-Up of Five Randomised Trials," *Lancet* 376 (2010): 1741–50.

91 In 2015, investigators from the Danish Cancer Society: S. Friis, A. H. Riis, R. Erichsen, et al., "Low-Dose Aspirin or Nonsteroidal Anti-Inflammatory Drug Use and Colorectal Cancer Risk: A Population-Based, Case-Control Study," *Annals of Internal Medicine* 163 (2015): 347–55.

91 Also in 2015, investigators . . . reviewed ten more studies: H. Nan, C. M. Hutter, E. J. Jacobs, et al., "Association of Aspirin and NSAID Use with Risk of Colorectal Cancer According to Genetic Variants," *Journal of the American Medical Association* 313 (2015): 1133–42.

91 how aspirin prevents colon cancer: A. T. Chan, S. Ogino, and C. S. Fuchs, "Aspirin and the Risk of Colorectal Cancer in Relation to Expression of COX-2," *New England Journal of Medicine* 350 (2007): 2131–42.

92 not all colon cancers make COX-2: Chan, Ogino, and Fuchs, "Aspirin and the Risk of Colorectal Cancer."

92 fatal bleeding events caused by aspirin: G. De Berardis, G. Lucisano, A. D'Ettorre, et al., "Association of Aspirin Use with Major Bleeding in Patients with and Without Diabetes," *Journal of the American Medical Association* 307 (2012): 2286–94; E. P. Whitlock, B. U. Burda, S. B. Williams, et al., "Bleeding Risks with Aspirin Use for Primary Prevention in Adults: A Systematic Review for the U.S. Preventive Services Task Force," *Annals of Internal Medicine* 164 (2016): 826–35; L. Li, O. C. Geraghty, A. Mehta, et al., "Age-Specific Risks, Severity, Time Course, and Outcome of Bleeding on Long-Term Antiplatelet Treatment After Vascular Events: A Population-Based Cohort Study," *Lancet* 390 (2017): 490–99.

Chapter 8: Embrace Allergenic Foods for Infants

97 a nurse in a San Francisco elementary: T. Sharples, "Have Americans Gone Nuts Over Nut Allergies?" *Time,* January 4, 2009.

97 Peanut phobia reached its illogical end: G. D. Braunstein, "Getting Past the Hype About Peanut Allergies," *Huffington Post*, May 18, 2013, https://www.huffingtonpost.com/glenn-d-braunstein-md/peanut-allergies_b_2885819.html.

97 admitted to the hospital with hives . . . death: S. M. Jones and A. W. Burks, "Food Allergy," *New England Journal of Medicine* 377 (2017): 1168–76.

98 "a Snickers bar 50 feet away is a lethal weapon": Quoted in Sharples, "Have Americans Gone Nuts Over Nut Allergies?"

99 recommended that women avoid eating peanuts during pregnancy: Jones and Burks, "Food Allergy."

99 introduction of peanuts into the diet: G. Du Toit, Y. Katz, P. Sasieni, et al., "Early Consumption of Peanuts in Infancy Is Associated with a Low Prevalence of Peanut Allergy," *Journal of Allergy and Clinical Immunology* 122 (2008): 984–91.

99 LEAP trial: G. Du Toit, G. Roberts, P. H. Sayre, et al., "Randomized Trial of Peanut Consumption in Infants at Risk for Peanut Allergy," *New England Journal of Medicine* 372 (2015): 803–13.

100 Four years of eating peanuts was enough: G. Du Toit, P. H. Sayre, G. Roberts, et al., "Effect of Avoidance of Peanut Allergy After Early Peanut Consumption," *New England Journal of Medicine* 374 (2016): 1435–43.

100 National Institutes of Health issued the following guidelines: A. Togias, S. F. Cooper, M. L. Acebal, et al., "Addendum Guidelines for the Prevention of Peanut Allergy in the United States: Report of the National Institute of Allergy and Infectious Diseases—Sponsored Expert Panel," *World Allergy Organization Journal* (2017), doi: 10.1186/s40413-016-0137-9.

101 mothers ate peanuts while breastfeeding: T. J. Pitt, A. B. Becker, M. Chan-Yeung, et al., "Reduced Rick of Peanut Sensitization Following Exposure Through Breast Feeding and Early Peanut Introduction," *Journal of Allergy and Clinical Immunology* 141 (2018): 620–25.

Chapter 9: The False Security of Sunblock

102 Skin cancer . . . die from it: R. P. Gallagher, "Sunscreens in Melanoma and Skin Cancer Prevention," *Canadian Medical Association Journal* 173 (2005): 244–45.

103 incidence of malignant melanoma is increasing: Anonymous, "Skin Cancer on the Rise," https://www.ewg.org/sunscreen/report/skin-cancer-on-the-rise/#.WyK/xq2-JuU.

103 Those at greatest risk: L. Dillner, "Does Sunscreen Prevent Skin Cancer?" *Guardian*, June 13, 2014.

103 ultraviolet (UV) radiation in three different wavelengths: A. R. Young, J. Claveau, and A. B. Rossi, "Ultraviolet Radiation and the Skin: Photobiology and Sunscreen Photoprotection," *Journal of the American Academy of Dermatology* 76 (2017): S100–9.

103 studies in experimental animals in the 1980s: P. D. Forbes, "Photocarcinogenesis: An Overview," *Journal of Investigative Dermatology* 77 (1981): 139–43; G. Kelfkens, H. van Weelden, F. R. de Gruijl, et al., "The Influence of Dose Rate on Ultraviolet Tumorigenesis," *Journal of Photochemistry and Photobiology* 10 (1991): 41–50; M. Situm, M. Buljan, S. O. Bulic, et al., "The Mechanisms of UV Radiation in the Development of Malignant Melanoma," *Collegium Antropologicum* 31 (2007): 13–16; K. Wischermann, S. Popp, S. Moshir, et al., "UVA Radiation Causes DNA Strand Breaks, Chromosomal Aberrations and Tumorigenic Transformation in HaCaT Skin Keratinocytes," *Oncogene* 27 (2008): 4269–80.

104 Tanning bed studies: B. Gerber, P. Mathys, M. Moser, et al., "Ultraviolet Emission Spectra of Sunbeds," *Photochemistry and Photobiology* 76 (2002): 664–68; International Agency for Research on Cancer, "The Association of Use of Sunbeds with Cutaneous Malignant Melanoma and Other Skin Cancers: A Systematic Review," *International Journal of Cancer* 120 (2006): 1116–22; D. Lazovich, R. K. Vogel, M. Berwick, et al., "Indoor Tanning and Risk of Melanoma: A Case-Control Study in a Highly Exposed Population," *Cancer Epidemiology, Biomarkers and Prevention* 19 (2010): 1557–68.

104 sun protection factor: Skin Cancer Foundation, "If Recent Attacks on Sunscreen Concern You . . . ," *Journal of the Dermatology Nurses' Association* 2 (2010): 228–29.

104 whether sunscreens of various SPFs prevent skin cancer: A. Green, G. Williams, R. Neale, et al., "Daily Sunscreen Application and Betacarotene Supplementation in Prevention of Basal-Cell and Squamous-Cell Carcinomas of the Skin: A Randomised Controlled Trial," *Lancet* 354 (1999): 723–29.

105 In 1999 . . . reviewed existing evidence: M. E. Bigby, "The Sunscreen and Melanoma Controversy," *Archives of Dermatology* 135 (1999): 1526–27.

106 In 2000, Autier and colleagues: P. Autier, J.-F. Doré, S. Négrier, et al., "Sunscreen Use and Duration of Sun Exposure: A Double-Blind, Randomized Trial," *Journal of the National Cancer Institute* 91 (1999): 1304–9; P. Autier, J.-F. Doré, A. C. Reis, et al., "Sunscreen Use and Intentional Exposure to Ultraviolet A and B Radiation: A Double Blind Randomized Trial Using Personal Dosimeters," *British Journal of Cancer* 83 (2000): 1243–48; P. Autier, M. Boniol, and J.-F. Doré, "Sunscreen Use and Increased Duration of Intentional Exposure: Still a Burning Issue," *International Journal of Cancer* 121 (2007): 1–5.

107 Marais studied 110 mice: A. Viros, B. Sanchez-Laorden, M. Pedersen, et al., "Ultraviolet Radiation Accelerates BRAF-Driven Melanomagenesis by Targeting TP53," *Nature* 511 (2014): 478–82.

107 "People tend to think they're invincible": B. Spencer, "Wearing Sunscreen May NOT Prevent Skin Cancer, Study Claims," *Daily Mail*, June 11, 2014.

108 The first line of defense . . . when UVB radiation is greatest: Skin Cancer Foundation, "If Recent Attacks on Sunscreen Concern You . . . "; M. B. Planta, "Sunscreen and Melanoma: Is Our Prevention Message Correct?" *Journal of the American Board of Family Medicine* 24 (2011): 735–39; D. Lazovich, R. I. Vogel, M. Berwick, et al., "Melanoma Risk in Relation to Use of Sunscreen or Other Sun Protection Methods," *Cancer Epidemiology, Biomarkers and Prevention* 20 (2011): 2583–93; Spencer, "Wearing Sunscreen May NOT Prevent Skin Cancer."

Chapter 10: Avoid Reflux Medicines for Fussy Babies

109 gastroesophageal reflux: J. R. Lightdale, D. A. Gremse, et al., "Gastroesophageal Reflux: Management Guidance for the Pediatrician," *Pediatrics* 131 (2013): e1684–95; H. A. Long, L. Solski, J. A. Rebuck, et al., "Infantile Gastroesophageal Reflux: Adherence to Treatment Guidelines in the Hospital Setting," *Journal of Pediatric Pharmacology and Therapeutics* 23 (2018): 41–47.

110 by the end of the first year: Lightdale, Gremse, et al., "Gastroesophageal Reflux."

110 symptoms that distinguish GER from GERD: Lightdale, Gremse, et al., "Gastroesophageal Reflux."

110 Two different types of medicine: P. T. Ramsey and A. Carr, "Gastric Acid and Digestive Physiology," *Surgical Clinics of North America* 91 (2011): 977–82.

111 In 2003, researchers from the University of Australia: D. J. Moore, B. S. Tao, D. R. Lines, et al., "Double-Blind Placebo Controlled Trial of Omeprazole in Irritable Infants with Gastroesophageal Reflux," *Journal of Pediatrics* 143 (2003): 219–23.

111 In 2009, researchers from Australia and Sweden: T. Omari, P. Lundborg, M. Sandström, et al., "Pharmacodynamics and Systemic Exposure of Esomeprazole in Preterm Infants and Term Neonates with Gastroesophageal Reflux Disease," *Journal of Pediatrics* 155 (2009): 222–28.

111 Also in 2009, researchers from the United States and Canada: S. R. Orenstein, E. Hassall, W. Furmaga-Jablonska, et al., "Multicenter, Double-Blind, Randomized, Placebo-Controlled Trial Assessing the Efficacy and Safety of Proton Pump Inhibitor Lansoprazole in Infants with Symptoms of Gastroesophageal Reflux Disease," *Journal of Pediatrics* 154 (2009): 514–20.

111 In 2010, researchers from the United States and Poland: H. Winter, P. Kum-Nji, S. H. Mahomedy, et al., "Efficacy and Safety of Pantoprazole Delayed-Release Granules for Oral Suspension in a Placebo-Controlled Treatment Withdrawal Study in Infants 1–11 Months Old with Symptomatic GERD," *Journal of Pediatric Gastroenterology and Nutrition* 50 (2010): 609–18.

112 where humans fit among other species: D. E. Beaseley, A. M. Koltz, J. E. Lambert, et al., "The Evolution of Stomach Acidity and Its Relevance to the Human Microbiome," *PLOS ONE* 10, no. 7 (2015): e0134116, doi: 10.1371/journal.pone.0134116.

113 greater risk of bacterial infections of the intestine: M. Safe, W. H. Chan, S. T. Leach, et al., "Widespread Use of Gastric Acid Inhibitors in Infants: Are They Needed? Are They Safe?" *World Journal of Gastrointestinal Pharmacology and Therapeutics* 7 (2016): 531–39.

113 Infants are also at greater risk: Safe, Chan, Leach, et al., "Widespread Use of Gastric Acid Inhibitors."

113 necrotizing enterocolitis: R. Guillet, B. J. Stoll, C. M. Cotton, et al., "Association of H2-Blocker Therapy and Higher Incidence of Necrotizing Enterocolitis in Very Low Birth Weight Infants," *Pediatrics* 117 (2006): e137–42; G. Terrin, A. Passariello, M. DeCurtis, et al., "Ranitidine Is Associated with Infections, Necrotizing Enterocolitis, and Fatal Outcome in Newborns," *Pediatrics* 129 (2012): e40–45.

113 increased risk of bone fractures: A. Malchodi, K. Wagner, A. Susi, et al., "Early Acid Suppression Therapy Exposure and Fracture in Young Children," *Pediatrics* 144 (2019): 10.

113 study of 800,000 children: E. Mitre, A. Susi, L. E. Knopp, et al., "Association Between Use of Acid-Suppressive Medications and Antibiotics During Infancy and Allergic Diseases in Early Childhood," *JAMA Pediatrics* 172 (2018), doi: 10.1001/jamapediatrics.2018.0315.

113 strategies . . . worth a try: Lightdale, Gremse, et al., "Gastroesophageal Reflux"; Anonymous, "Medication for Infant Reflux? Why You Should Probably Pass," April 9, 2018, https://www.kidnurse.org/infant-reflux.

114 American Academy of Pediatrics stated: M. Boruta, R. Boruta, and J. Li, "Acid Suppression Therapy and Symptom Improvement (or Lack Thereof) in Children," *Pediatrics* 144 (2019): 11.

114 one out of every ten babies: Malchodi, Wagner, Susi, et al., "Early Acid Suppression Therapy Exposure."

Chapter 11: Prostate Cancer Screening Programs Do More Harm Than Good

117 a barn containing three animals: A. Gawande, "Overkill: An Avalanche of Unnecessary Medical Care Is Harming Patients Physically and Financially. What Can We Do About It?" *New Yorker*, May 11, 2015.

118 "We now have a vast and costly healthcare industry": Gawande, "Overkill."

118 one out of every seven men: V. A. Moyer, on behalf of the U.S. Preventive Services Task Force, "Screening for Prostate Cancer: U.S. Preventive Services Task Force Recommendation Statement," *Annals of Internal Medicine* 157 (2012): 120–34; E. A. Miller, P. F. Pinsky, A. Black, et al., "Secondary Prostate Cancer Screening Outcomes by

Race in the Prostate, Lung, Colorectal, and Ovarian (PLCO) Screening Trial," *The Prostate* 78 (2018): 830–38; L. Esserman, Y. Shieh, and I. Thompson, "Rethinking Screening for Breast Cancer and Prostate Cancer," *Journal of the American Medical Association* 302 (2009): 1685–92; J. Eastham, "Prostate Cancer Screening," *Investigative and Clinical Urology* 58 (2017): 217–19; D. Grady, "Early Prostate Cancer Cases Fall Along with Screening," *New York Times*, November 17, 2015.

119 Food and Drug Administration approved PSA levels: Eastham, "Prostate Cancer Screening."

119 no one popularized PSA tests more than Bob Dole: M. McCullough, "PSA Screening Events on the Ropes—But Not Gone," *Philadelphia Inquirer*, December 16, 2016.

119 (PLCO) Trial: P. F. Pinsky, P. C. Prorok, K. Yu, et al., "Extended Mortality Results for Prostate Cancer Screening in the PLCO Trial with Median Follow-up of 15 Years," *Cancer* 123 (2017): 592–99; H. J. de Koning, R. Gulati, S. M. Moss, et al., "The Efficacy of Prostate-Specific Antigen Screening: Impact of Key Components in the ERSPC and PLCO Trials," *Cancer* 124 (2018): 1197–206.

120 ERSPC: F. H. Schröder, J. Hugosson, M. J. Roobol, et al., "Screening and Prostate Mortality Results of the European Randomised Study of Screening for Prostate Cancer (ERSPC) at 13 Years Follow-up," *Lancet* 384 (2014): 2027–35; de Koning, Gulati, Moss, et al., "The Efficacy of Prostate-Specific Antigen Screening."

120 CAP: R. M. Martin, J. L. Donovan, E. L. Turner, et al., "Effect of Low-Intensity PSA-Based Screening Intervention on Prostate Cancer Mortality: The CAP Randomized Clinical Trial," *Journal of the American Medical Association* 319 (2018): 883–95.

120 ProtecT: P. C. Hamdy, J. L. Donovan, J. A. Lane, et al., "10-Year Outcomes After Monitoring, Surgery, or Radiotherapy for Localized Prostate Cancer," *New England Journal of Medicine* 375 (2016): 1415–24.

121 deciding whether to enter a prostate screening program: Moyer, on behalf of the U.S. Preventive Services Task Force, "Screening for Prostate Cancer"; R. M. Hoffman, "Implications of the New USPSTF Prostate Cancer Screening Recommendation—Attaining Equipoise," *JAMA Internal Medicine* 178 (2018): 889–91.

121 Prostate biopsies are also a rare cause of bacterial infections of the bloodstream: H. Wolinsky, "Men, Beware of Biopsies for Prostate Cancer Dx," *Medpage Today*, August 19, 2018.

122 For every 1,000 men screened: A. Buelt, "Questioning Medicine: Prostate Cancer Screening," *Medpage Today*, October 5, 2014; M. McCullough, "Is Prostate Cancer Screening Right for You? Let's Do the Numbers," *Philadelphia Inquirer*, May 8, 2018.

122 As a consequence of these findings: P. F. Pinsky, P. C. Prorok, and B. S. Kramer, "Prostate Cancer Screening—A Perspective on the Current State of the Evidence," *New England Journal of Medicine* 376 (2017): 1285–89.

122 number of men entering prostate cancer screening programs has been cut in half: McCullough, "PSA Screening Events on the Ropes—But Not Gone."

122 "I never dreamed that my discovery": R. J. Ablin, "The Great Prostate Mistake," *New York Times*, March 9, 2010.

123 isn't there something that can be done?: D. Gorski, "The Cancer Screening Kerfuffle Erupts Again: 'Rethinking' Screening for Breast and Prostate Cancer," *Science Based Medicine*, November 2, 2009.

Chapter 12: Thyroid Cancer Screening Programs Don't Save Lives

124 incidence of thyroid cancer in the United States: U.S. Preventive Services Task Force, "Screening for Thyroid Cancer: U.S. Preventive Services Task Force Recommendation," *Journal of the American Medical Association* 317 (2017): 1882–87.

124 France, Italy, Croatia, Israel, China, Australia, Switzerland, Canada, and the Czech Republic: H. S. Ahn, J. J. Kim, and H. G. Welch, "Korea's Thyroid-Cancer 'Epidemic'—Screening and Overdiagnosis," *New England Journal of Medicine* 371 (2014): 1765–67; S. Jegerlehner, J.-L. Bulliard, D. Aujesky, et al., "Overdiagnosis and Overtreatment of Thyroid Cancer: A Population-Based Temporal Trend Study," *PLOS One*, doi: 10.1371/journalpone.0179387.

125 incidence of thyroid cancer in South Korea: Ahn, Kim, and Welch, "Korea's Thyroid-Cancer 'Epidemic.'"

125 "tsunami of thyroid cancer": G. Kolata, "Study Points to Overdiagnosis of Thyroid Cancer," *New York Times*, November 5, 2014.

125 exposure to radiation: S. M. Nielsen, M. G. White, S. Hong, et al., "The Breast-Thyroid Cancer Link: A Systematic Review and Meta-Analysis," *Cancer Epidemiology, Biomarkers and Prevention* 25 (2016): 231–38.

125 Radiation exposure . . . cannot explain the current epidemic: "United States Nuclear Tests: July 1945 Through September 1992," DOE/NV—209-REV 15, December 2000, Las Vegas: Department of Energy, Nevada Operations Office.

126 radiation to "treat" common disorders: L. Davies and H. G. Welch, "Increasing Incidence of Thyroid Cancer in the United States, 1973–2002," *Journal of the American Medical Association* 295 (2006): 2164–67.

126 "Check Your Neck": S. Rochman, "Thyroid Cancer's Overdiagnosis Problem," *Journal of the National Cancer Institute* (2017), doi: 10.1093/jnci/djx153.

126 quickly adopted for widespread use: Davies and Welch, "Increasing Incidence of Thyroid Cancer."

127 "fine needle biopsy": Davies and Welch, "Increasing Incidence of Thyroid Cancer."

127 total removal of the thyroid: H. G. Welch and G. M. Doherty, "Saving Thyroids—Overtreatment of Small Papillary Cancers," *New England Journal of Medicine* 379 (2018): 310–12.

127 vocal cord paralysis: J. S. Lin, E. J. Aiello Bowles, S. B. Williams, et al., "Screening for Thyroid Cancer: Update Evidence Report and Systematic Review for the US Preventive Services Task Force," *Journal of the American Medical Association* 317 (2017): 1888–903.

127 abnormal levels of calcium in the bloodstream and bones: Lin, Bowles, Williams, et al., "Screening for Thyroid Cancer."

127 radioactive iodine: L. Davies and H. G. Welch, "Current Thyroid Cancer Trends in the United States," *JAMA Otolaryngology—Head and Neck Surgery* 140 (2014): 317–22.

127 need to replace the hormone produced by the thyroid: J. Jonklass, A. C. Bianco, A. J. Bauer, et al., "Guidelines for the Treatment of Hy-

pothyroidism: Prepared by the American Thyroid Association Task Force on Thyroid Hormone Replacement," *Thyroid* 24 (2014): 1670–751.

128 therapies haven't saved lives: Welch and Doherty, "Saving Thyroids."

128 "epidemic of diagnosis": Kolata, "Study Points to Overdiagnosis of Thyroid Cancer."

128 survival rate . . . statistically indistinguishable: L. Davies and H. G. Welch, "Thyroid Cancer Survival in the United States: Observational Data from 1973 to 2005," *JAMA Otolaryngology—Head and Neck Surgery* 136 (2010): 440–44.

128 papillary cancer: Davies and Welch, "Increasing Incidence of Thyroid Cancer."

129 researchers in Finland examined the thyroids of 101 people: H. R. Harach, K. O. Franssila, and V.-M. Wasenius, "Occult Papillary Carcinoma of the Thyroid: A 'Normal' Finding in Finland. A Systematic Autopsy Study," *Cancer* 56 (1985): 531–38.

129 Japanese researchers later extended the Finnish study: T. Takano, "Natural History of Thyroid Cancer," *Endocrine Journal* 64 (2017): 237–44.

129 "If it's not cancer": G. Kolata, "It's Not Cancer: Doctors Reclassify a Thyroid Tumor," *New York Times*, April 14, 2016.

129 "classified as a normal finding": Davies and Welch, "Increasing Incidence of Thyroid Cancer."

129 advisory bodies . . . recommend against thyroid cancer screening: U.S. Preventive Services Task Force, "Screening for Thyroid Cancer."

130 "The message for so long": Kolata, "Study Points to Overdiagnosis of Thyroid Cancer."

130 "Don't do anything": Kolata, "Study Points to Overdiagnosis of Thyroid Cancer."

130 "You don't make a lot of money doing ultrasounds and biopsies": R. M. Tuttle and G. Miller, "Overdiagnosis in Thyroid Cancer: The Case for Observation," *Medscape*, August 25, 2015.

131 "I saw someone today whose hair was getting thin": Tuttle and Miller, "Overdiagnosis in Thyroid Cancer."

131 Physician Coalition for Prevention of Overdiagnosis of Thyroid Cancer: H. S. Ahn and H. G. Welch, "South Korea's Thyroid-Cancer 'Epidemic'—Turning the Tide," *New England Journal of Medicine* 373 (2015): 2389–90; Ahn, Kim, and Welch, "Korea's Thyroid-Cancer 'Epidemic.'"

Chapter 13: Breast Cancer Screening Programs Aren't Exactly as Advertised

133 Breast cancer . . . diagnose it early: J. S. Mandelblatt, K. A. Cronin, S. Bailey, et al., "Effects of Mammography Screening Under Different Screening Schedules: Model Estimates of Potential Benefits and Harms," *Annals of Internal Medicine* 151 (2009): 738–47; H. D. Nelson, K. Tyne, A. Naik, et al., "Screening for Breast Cancer: An Update for the U.S. Preventive Services Task Force," *Annals of Internal Medicine* 151 (2009): 727–37; R. Shah, K. Rosso, and D. Nathanson, "Pathogenesis, Prevention, Diagnosis and Treatment of Breast Cancer," *World Journal of Clinical Oncology* 5 (2014): 283–98; World Health Organization, "WHO Position Paper on Mammography Screening," Geneva, World Health Organization 2014; K. C. Oeffinger, E. T. H. Fontham, R. Etzioni, et al., "Breast Cancer Screening for Women at Average Risk, 2015 Guideline Update from the American Cancer Society," *Journal of the American Medical Association* 314 (2015): 1559–614; A. L. Siu, "Screening for Breast Cancer: U.S. Preventive Services Task Force Recommendation Statement," *Annals of Internal Medicine* 164 (2016): 279–96; Y.-S. Sun, Z. Zhao, A.-N. Yang, et al., "Risk Factors and Preventions of Breast Cancer," *International Journal of Biological Science* 13 (2017): 1387–97.

134 Röntgen . . . Egan: L. H. Baker, "Breast Cancer Detection Demonstration Project: Five-Year Summary Report," *CA: A Cancer Journal for Clinicians* 32 (1982): 194–225.

134 "The present communication offers the first hard evidence": S. Shapiro, P. Strax, and L. Venet, "Periodic Breast Cancer Screening in Reducing Mortality from Breast Cancer," *Journal of the American Medical Association* 215 (1971): 1777–85.

135 In 1985, researchers in Sweden: L. Tabár, C. J. Fagerberg, A. Gad, et al., "Reduction in Mortality from Breast Cancer After Mass Screening with Mammography: Randomised Trial from the Breast Cancer Screening Working Group of the Swedish National Board of Health and Welfare," *Lancet* 1 (1985): 829–32.

135 In 1993, researchers . . . reviewed the results of eight studies: S. W. Fletcher, W. Black, R. Harris, et al., "Report of the International Workshop on Screening for Breast Cancer," *Journal of the National Cancer Institute* 85 (1993): 1644–56.

135 deaths from breast cancer steadily decreased: Nelson, Tyne, Naik, et al., "Screening for Breast Cancer."

135 In 2000, researchers . . . reviewed the results of eight studies: P. C. Gøtzsche and O. Olsen, "Is Screening for Breast Cancer with Mammography Justifiable?" *Lancet* 355 (2000): 129–34; O. Olsen and P. C. Gøtzsche, "Screening for Breast Cancer with Mammography (A Review)," *Cochrane Database of Systematic Reviews* 4 (2001): CD001877.

135 "The trials with adequate randomization": P. C. Gøtzsche and K. J. Jørgensen, "Screening for Breast Cancer with Mammography (Review)," *Cochrane Database of Systematic Reviews* 6 (2013): CD001877.

136 "Annual mammography in women": A. B. Miller, C. Wall, C. J. Baines, et al., "Twenty-Five Year Follow-Up for Breast Cancer Incidence and Mortality of the Canadian Breast Screening Study: Randomised Screening Trial," *British Medical Journal* 348 (2014): g366.

136 During the past twenty years, advances: A. G. Waks and E. P. Winer, "Breast Cancer Treatment: A Review," *Journal of the American Medical Association* 321 (2019): 288–300.

137 The impact of breast cancer treatments: World Health Organization, "WHO Position Paper on Mammography Screening."

137 In 2012, Archie Bleyer: A. Bleyer and H. G. Welch, "Effect of Three Decades of Screening Mammography on Breast-Cancer Incidence," *New England Journal of Medicine* 367 (2012): 1998–2005.

138 physical and emotional cost of early detection: S. C. Darby, P. McGale, M. Ewertz, et al., "Risk of Ischemic Heart Disease in Women After Radiotherapy for Breast Cancer," *New England Journal of Medicine* 368 (2013): 987–98; P. C. Gøtzsche, "Mammography Screening Is Harmful and Should Be Abandoned," *Journal of the Royal Society of Medicine* 108 (2015): 341–45.

139 "Overdiagnosis and unnecessary treatment of nonfatal cancer": M. Kalager, H. O. Adami, M. Bretthauer, et al., "Overdiagnosis of Invasive Breast Cancer Due to Mammography Screening: Results from

the Norwegian Screening Program," *Annals of Internal Medicine* 156 (2012): 491–99.

139 "physicians have been more likely to discuss benefits than harms": M. S. Fuller, C. I. Lee, and J. G. Elmore, "Breast Cancer Screening: An Evidence-Based Update," *Medical Clinics of North America* 99 (2015): 451–68.

139 "most patient-education ads . . . informed choice": A. Bleyer, "Screening Mammography: Update and Review of Publications Since Our Report in the *New England Journal of Medicine* on the Magnitude of the Problem in the United States," *Academic Radiology* 22 (2015): 949–60.

140 In 1996, researchers from Stockholm: E. Lidbrink, J. Elfving, J. Frisell, et al., "Neglected Aspects of False Positive Findings of Mammography and Breast Cancer Screening: Analysis of False Positive Cases from Stockholm Trial," *British Medical Journal* 312 (1996): 273.

140 In 1998, researchers from the University of Washington: J. G. Elmore, M. B. Barton, V. M. Moceri, et al., "Ten-Year Risk of False Positive Screening Mammograms and Clinical Breast Examinations," *New England Journal of Medicine* 338 (1998): 1089–96.

141 In 2012, researchers from Oslo: S. Hofvind, A. Ponti, J. Patnick, et al., "False-Positive Results in Mammographic Screening for Breast Cancer in Europe: A Literature Review and Survey of Service Screening Programmes," *Journal of Medical Screening* 19 Suppl. 1 (2012): 57–66.

141 For every 1,000 women screened every two years: Gøtzsche and Olsen, "Is Screening for Breast Cancer with Mammography Justifiable?"; Mandelblatt, Cronin, Bailey, et al., "Effects of Mammography Screening Under Different Screening Schedules"; Independent UK Panel on Breast Cancer Screening, "The Benefits and Harms of Breast Cancer Screening: An Independent Review," *Lancet* 380 (2012): 1778–86; EUROSCREEN Working Group, "Summary of the Evidence of Breast Cancer Service Screening Outcomes in Europe and First Estimate of the Benefit and Harm Balance Sheet," *Journal of Medical Screening* 19 Suppl 1 (2012): 5–13; Gøtzsche and Jørgensen, "Screening for Breast Cancer with Mammography (Review)"; M. Løberg, M. L. Lousdal, M. Bretthauer, et al., "Benefits and Harms of Mammography Screening," *Breast Cancer Research* 17 (2015): 63;

S. M. Moss, C. Wale, R. Smith, et al., "Effect of Mammographic Screening from Age 40 Years on Breast Cancer Mortality in the UK Age Trial 17 Years' Follow-up: A Randomised Controlled Trial," *Lancet* 16 (2015): 1123–32; A. B. Nattinger and J. L. Mitchell, "Breast Cancer Screening and Prevention," *Annals of Internal Medicine* 164 (2016): ITC81–96; Siu, "Screening for Breast Cancer."

141 recommendations from various advisory groups: Oeffinger, Fontham, Etzioni, et al., "Breast Cancer Screening for Women at Average Risk"; Siu, "Screening for Breast Cancer"; S. Raichand, A. G. Dunn, M.-S. Ong, et al., "Conclusions in Systematic Reviews of Mammography for Breast Cancer Screening and Associations with Review Design and Author Characteristics," *Systematic Reviews* 6 (2017): 105.

142 69 percent of advertisements for cancer prevention: T. J. Caverly, R. A. Hayward, E. Reamer, et al., "Presentation of Benefits and Harms in US Cancer Screening and Prevention Guidelines: Systematic Review," *Journal of the National Cancer Institute* 108 (2016): djv436.

143 ductal carcinoma in situ: C. M. Vatovec, M. Z. Erten, J. Kolodinsky, et al., "Ductal Carcinoma *in Situ*: A Brief Review of Treatment Variation and Impacts on Patients and Society," *Critical Reviews of Eukaryotic Gene Expression* 24 (2014): 281–86; K. L. Gorringe and S. B. Fox, "Ductal Carcinoma In Situ Biology, Biomarkers, and Diagnosis," *Frontiers in Oncology* 7 (2017): 248.

144 Big Bet phenomenon was tested years ago at the racetrack: R. E. Knox and J. A. Inkster, "Postdecision Dissonance at Post Time," *Journal of Personality and Social Psychology* 8 (1968): 319–23.

Chapter 14: Heart Stents Don't Prolong Lives

147 Heart disease is the leading cause of death worldwide: E. G. Nabel and E. Braunwald, "A Tale of Coronary Artery Disease and Myocardial Infarction," *New England Journal of Medicine* 366 (2012): 54–63.

147 By 2030, the worldwide death rate from heart disease: A. Cassar, D. R. Holmes, C. S. Rihal, et al., "Chronic Coronary Artery Disease: Diagnosis and Management," *Mayo Clinic Proceedings* 84 (2009): 1130–46.

148 In the United States, 17 million people suffer from heart disease: Cassar, Holmes, Rihal, et al., "Chronic Coronary Artery Disease."

148 treated with bloodletting, opium, and bed rest: Nabel and Braunwald, "A Tale of Coronary Artery Disease."

148 The mortality rate in people admitted to the hospital following a heart attack: Nabel and Braunwald, "A Tale of Coronary Artery Disease."

149 habits of five thousand people living in Framingham, Massachusetts: W. B. Kannel, T. R. Dawber, A. Kagan, et al., "Factors of Risk in the Development of Coronary Artery Disease—Six Year Follow-up Experience: The Framingham Study," *Annals of Internal Medicine* 55 (1961): 33–50.

149 Ancel Keys . . . trans fats: G. Taubes, "The Soft Science of Dietary Fat," *Science* 291 (2001): 2536–45.

150 Surgeon General . . . concluded that cigarette smoking: U.S. Surgeon General's Advisory Committee on Smoking and Health, "Smoking and Health," Washington, DC: U.S. Public Health Service, Office of the Surgeon General, 1964 (PHS publication no. 1103).

150 Werner Forssmann: W. Forssmann, "Catheterization of the Right Heart," *Klinische Wochenschrift* 8 (1929): 2085–87.

150 inject a dye into coronary arteries: F. M. Sones Jr. and E. K. Shirey, "Cine Coronary Arteriography," *Modern Concepts of Cardiovascular Disease* 31 (1962): 735.

151 trial of more than ten thousand people: Gruppo Italiano per lo Studio della Streptochinasi nell'Infarto Miocardico (GISSI), "Effectiveness of Intravenous Thrombolytic Treatment in Acute Myocardial Infarction," *Lancet* 327 (1986): 397–402.

151 In 1992, researchers from the Veterans Affairs Medical Centers: A. F. Parisi, E. D. Folland, and P. Hartigan, "A Comparison of Angioplasty with Medical Therapy in the Treatment of Single-Vessel Coronary Artery Disease," *New England Journal of Medicine* 326 (1992): 10–16.

152 In 2005, researchers from the Netherlands: R. J. De Winter, F. Windhausen, J. H. Cornel, et al., "Early Invasive Versus Selectively Invasive Management for Acute Coronary Syndromes," *New England Journal of Medicine* 353 (2005): 1095–104.

152 Also in 2005, researchers in Greece: D. G. Katritsis and J. P. A. Ioannidis, "Percutaneous Coronary Intervention Versus Conservative Therapy in Nonacute Coronary Artery Disease," *Circulation* 111 (2005): 2906–12.

153 In 2006, researchers from New York University: J. S. Hochman, G. A. Lamas, C. E. Buller, et al., "Coronary Intervention for Persistent Occlusion After Myocardial Infarction," *New England Journal of Medicine* 355 (2006): 2395–407.

153 In 2007, researchers from fifty U.S. and Canadian medical centers: W. E. Boden, R. A. O'Rourke, K. K. Teo, et al., "Optimal Medical Therapy With or Without PCI for Stable Coronary Disease," *New England Journal of Medicine* 356 (2007): 1503–6.

153 A follow-up of this study: W. S. Weintraub, J. A. Spertus, P. Kolm, et al., "Effect of PCI on Quality of Life in Patients with Stable Coronary Disease," *New England Journal of Medicine* 359 (2008): 677–87.

153 In 2009, researchers . . . reviewed twenty years of studies: T. A. Trikalinos, A. A. Alsheikh-Ali, A. Tatsioni, et al., "Percutaneous Coronary Interventions for Non-Acute Coronary Artery Disease: A Quantitative 20-Year Synopsis and a Network Meta-Analysis," *Lancet* 373 (2009): 911–18.

153 In 2014, researchers . . . reviewed yet another five studies: K. Stergiopoulos, W. E. Boden, P. Hartigan, et al., "Percutaneous Coronary Intervention Outcomes in Patients with Stable Obstructive Coronary Artery Disease and Myocardial Ischemia: A Collaborative Meta-Analysis of Contemporary Randomized Clinical Trials," *JAMA Internal Medicine* 174 (2014): 232–40.

154 Al-Lamee . . . controversial study: R. Al-Lamee, D. Thompson, H.-M. Dehbi, et al., "Percutaneous Coronary Intervention in Stable Angina (ORBITA): A Double-Blind, Randomised Controlled Trial," *Lancet* 391 (2018): 31–40.

154 stenting was easy to understand: Stergiopoulos, Boden, Hartigan, et al., "Percutaneous Coronary Intervention Outcomes."

155 "This should make us step back and ask questions": J. Belluz, "Thousands of Heart Patients Get Stents That May Do More Harm Than Good," Vox, November 6, 2017.

Chapter 15: Surgery for Knee Arthritis Is Unnecessary

156 twenty million Americans experience severe knee pain: S. T. Skou, E. M. Roos, M. B. Laursen, et al., "Total Knee Replacement and Non-Surgical Treatment of Knee Osteoarthritis: 2-Year Outcome

from Two Parallel Randomized Controlled Trials," *Osteoarthritis and Cartilage* 26 (2018): 1170–80; R. A. Siemieniuk, I. A. Harris, T. Agoritsas, et al., "Arthroscopic Surgery for Degenerative Knee Arthritis and Meniscal Tears: A Clinical Practice Guideline," *British Medical Journal* 357 (2017): j1982; R. Brignardello-Peterson, G. H. Guyatt, R. Buchbinder, et al., "Knee Arthroscopy Versus Conservative Management in Patients with Degenerative Knee Disease: A Systematic Review," *British Medical Journal*, doi: 10.1136/bmjopen-2017-016114; D. Bhatia, T. Bejarano, and M. Novo, "Current Interventions in the Management of Knee Osteoarthritis," *Journal of Pharmacy and Bioallied Sciences* 5 (2013): 30–38; A. M. Wood, T. M. Brock, K. Heil, et al., "A Review of the Management of Hip and Knee Osteoarthritis," *International Journal of Chronic Diseases*, doi: 10.1155/2013/845015; C. T. Hepper, J. J. Halvorson, S. T. Duncan, et al., "The Efficacy and Duration of Intra-Articular Corticosteroid Injection for Knee Osteoarthritis: A Systematic Review of Level I Studies," *Journal of the American Academy of Orthopedic Surgeons* 17 (2009): 638–46; D. O. Clegg, D. J. Reda, C. L. Harris, et al., "Glucosamine, Chondroitin Sulfate, and the Two in Combination for Painful Knee Arthritis," *New England Journal of Medicine* 354 (2006): 795–808; D. T. Felson, "Osteoarthritis of the Knee," *New England Journal of Medicine* 354 (2006): 841–48; J. H. Liang and P. Fortin, "Management of Osteoarthritis of the Hip and Knee," *New England Journal of Medicine* 325 (1991): 125–27.

157 No joint bears a larger burden than the knee: Bhatia, Bejarano, and Novo, "Current Interventions in the Management of Knee Osteoarthritis"; Felson, "Osteoarthritis of the Knee"; D. Hamerman, "The Biology of Osteoarthritis," *New England Journal of Medicine* 320 (1989): 1322–30.

158 In 2006, researchers from Germany: H.-P. Scharf, U. Mansmann, K. Streitberger, et al., "Acupuncture and Knee Arthritis: A Three-Armed Randomized Trial," *Annals of Internal Medicine* 145 (2006): 12–20.

159 Other anti-inflammatory drugs: Felson, "Osteoarthritis of the Knee."

159 injecting steroids into the knee joint: Hepper, Halvorson, Duncan, et al., "The Efficacy and Duration of Intra-Articular Corticosteroid Injection for Knee Osteoarthritis"; P. A. Dieppe, B. Sathapatayavongs, J. E. Jones, et al., "Intra-Articular Steroids in Osteoarthritis," *Rheumatology and Rehabilitation* 19 (1980): 212–17.

159 steroids are . . . not a long-term solution: G. N. Chandler, D. M. Oxon, and V. Wright, "Deleterious Effect of Intra-Articular Hydrocortisone," *Lancet* 262 (1958): 661–63; C. Steinburg, R. B. Duthie, and A. E. Piva, "Charcot-Like Arthropathy Following Intra-Articular Hydrocortisone," *Journal of the American Medical Association* 181 (1962): 851–54; A. M. Lufti and K. Kosel, "Effects of Intra-articularly Administered Corticosteroids and Salicylates on the Surface Structure of Cartilage," *Journal of Anatomy* 127 (1978): 393–402.

160 arthroscopic surgery offered substantial pain relief: J. B. Moseley, K. O'Malley, N. J. Petersen, et al., "A Controlled Trial of Arthroscopic Surgery for Osteoarthritis of the Knee," *New England Journal of Medicine* 347 (2002): 81–88.

160 In 2002, researchers . . . unethical study: Moseley, O'Malley, Petersen, et al., "A Controlled Trial of Arthroscopic Surgery for Osteoarthritis of the Knee."

161 "Although smoothing cartilage and meniscal irregularities may sound appealing": D. T. Felson and J. Buckwalter, "Débridement and Lavage for Osteoarthritis of the Knee," *New England Journal of Medicine* 347 (2002): 132–33.

162 In 2007, researchers from Sweden: S. Herrlin, M. Hallander, P. Wange, et al., "Arthroscopic or Conservative Treatment of Degenerative Medial Meniscal Tears: A Prospective Randomised Trial," *Knee Surgery Sports Traumatology Arthroscopy* 15 (2007): 393–401.

162 In 2008, researchers from Canada: A. Kirkley, T. B. Birmingham, R. B. Litchfield, et al., "A Randomized Trial of Arthroscopic Surgery for Osteoarthritis of the Knee," *New England Journal of Medicine* 359 (2008): 1097–107.

162 In 2013, researchers from Sweden: S. V. Herrlin, P. O. Wange, G. Lapidus, et al., "Is Arthroscopic Surgery Beneficial in Treating Non-Traumatic, Degenerative Medial Meniscus Tears? A Five-Year Follow-Up," *Knee Surgery Sports Traumatology Arthroscopy* 21 (2013): 358–64.

163 Also in 2013, researchers at Brigham and Women's Hospital: J. N. Katz, R. H. Brophy, C. E. Chaisson, et al., "Surgery Versus Physical Therapy for a Meniscal Tear and Osteoarthritis," *New England Journal of Medicine* 368 (2013): 1675–84.

163 In 2018, researchers in the Netherlands: V. A. van de Graaf, J. C. A. Noorduyn, M. W. Willigenburg, et al., "Effect of Early Surgery vs. Physical Therapy on Knee Function Among Patients with Nonobstructive Meniscal Tears: The ESCAPE Randomized Clinical Trial," *Journal of the American Medical Association* 320 (2018): 1328–37.

163 fourteen randomized, controlled clinical trials and twelve observational studies: Siemieniuk, Harris, Agoritsas, et al., "Arthroscopic Surgery for Degenerative Knee Arthritis."

164 In 2006, researchers from the University of Utah School of Medicine: Clegg, Reda, Harris, et al., "Glucosamine, Chondroitin Sulfate."

165 they were called parachute studies: J. Katz, "Parachutes and Preferences—A Trial of Knee Replacement," *New England Journal of Medicine* 373 (2015): 1668–69.

165 In 2015, researchers from Denmark: S. T. Skou, E. M. Roos, M. B. Laursen, et al., "A Randomized, Controlled Trial of Total Knee Replacement," *New England Journal of Medicine* 373 (2015): 1597–606.

166 In 2018, these same researchers: S. T. Skou, E. M. Roos, O. Simonsen, et al., "The Effects of Total Knee Replacement and Non-Surgical Treatment on Pain Sensitization and Clinical Pain," *European Journal of Pain* 20 (2016): 1612–21; Skou, Roos, Laursen, et al., "Total Knee Replacement and Non-Surgical Treatment of Knee Osteoarthritis."

166 physical therapy: Bhatia, Bejarano, and Novo, "Current Interventions in the Management of Knee Osteoarthritis"; Felson, "Osteoarthritis of the Knee."

Chapter 16: Don't Remove Mercury Dental Fillings

168 To treat cavities: National Capital Poison Control, "Do Fillings Cause Mercury Poisoning," https://www.poison.org/articles/2010-dec/do-fillings-cause-mercury-poisoning; T. G. Berry, J. B. Summitt, A. K. H. Chung, et al., "Amalgam at the New Millennium," *Journal of the American Dental Association* 129 (1998): 1547–56.

169 "the amalgam wars": M. Rathore, A. Singh, and V. A. Pant, "The Dental Amalgam Toxicity Fear: A Myth or Actuality," *Toxicology International* 19 (2012): 81–88; J. E. Dodes, "The Amalgam Controversy:

An Evidence-Based Analysis," *Journal of the American Dental Association* 132 (2001): 348–56.

169 Hal A. Huggins . . . still seeing patients: National Capital Poison Control, "Do Fillings Cause Mercury Poisoning"; S. Barrett, "The 'Mercury Toxicity' Scam: How Anti-Amalgamists Swindle People," Quackwatch, www.quackwatch.org/01QuackeryRelatedTopic/mercury.html.

170 Minamata: M. Harada, "Minamata Disease: Methylmercury Poisoning in Japan Caused by Environmental Pollution," *Critical Reviews in Toxicology* 25 (1995): 1–24; S. Juan, "The Minamata Disaster—50 Years On," *The Register,* June 14, 2006.

171 93,000 tons of wheat and barley . . . methylmercury, a fungicide: R. E. Fermer, "Chemical Disaster," *Pharmacology and Therapeutics* 58 (1993): 157–71.

171 Everyone living on the planet is exposed to mercury: National Capital Poison Control, "Do Fillings Cause Mercury Poisoning"; J. Chey and S. Buchanan, "Toxins in Everyday Life," *Primary Care: Clinics in Office Practice* 35 (2008): 707–27; Dodes, "The Amalgam Controversy."

173 nitrous mercury vapor: J. R. Mackert Jr., "Factors Affecting Estimation of Dental Amalgam Mercury Exposure from Measurements of Mercury Vapor Levels in Intra-Oral and Expired Air," *Journal of Dental Research* 66 (1987): 1775–80; S. Olsson and M. Bergman, "Daily Dose Calculations from Measurements of Intra-Oral Mercury Vapor," *Journal of Dental Research* 71 (1992): 414–23; J. R. Mackert Jr. and A. Berglund, "Mercury Exposure from Dental Amalgam Fillings: Absorbed Dose and the Potential for Adverse Health Effects," *Critical Review of Oral Biology and Medicine* 8 (1997): 410–36; Dodes, "The Amalgam Controversy"; S. A. Counter and L. H. Buchanan, "Mercury Exposure in Children: A Review," *Toxicology and Applied Pharmacology* 198 (2004): 209–30.

173 In 1988, investigators from Sweden: M. Ahlqwist, C. Bengtsson, B. Furunes, et al., "Number of Amalgam Tooth Fillings in Relation to Subjectively Experienced Symptoms in a Study of Swedish Women," *Community Dental Oral Epidemiology* 16 (1988): 227–31.

176 *60 Minutes* program: National Capital Poison Control, "Do Fillings Cause Mercury Poisoning"; Barrett, "The 'Mercury Toxicity' Scam"; Dodes, "The Amalgam Controversy."

176 *Consumer Reports* published a letter: Barrett, "The 'Mercury Toxicity' Scam"; S. Barrett, *Health Schemes, Scams, and Frauds* (Mount Vernon, NY: Consumer Reports Books, 1990).

176 "the greatest unfounded health scares of modern times": Barrett, "The 'Mercury Toxicity' Scam."

177 In 1991, researchers from the University of Kentucky: S. R. Saxe, D. A. Snowden, M. W. Wekstein, et al., "Dental Amalgam and Cognitive Function in Older Women: Findings from the Nun Study," *Journal of the American Dental Association* 126 (1995): 1495–501.

177 In 2005, researchers from the National Institutes of Health: A. Kingman, J. W. Albers, J. C. Arezzo, et al., "Amalgam Exposure and Neurological Function," *Neurotoxicity* 26 (2005): 241–55.

177 In 1998, researchers from the University of Montreal: D. Bangsi, P. Ghadirian, S. Ducic, et al., "Dental Amalgam and Multiple Sclerosis: A Case-Control Study in Montreal, Canada," *International Journal of Epidemiology* 27 (1998): 667–71.

177 In 2004, researchers from New Zealand: M. N. Bates, J. Fawcett, N. Garrett, et al., "Health Effects of Dental Amalgam Exposure: A Retrospective Cohort Study," *International Journal of Epidemiology* 33 (2004): 894–902.

178 "no significant association of Alzheimer's": S. R. Saxe, M. W. Wekstein, R. J. Kryscio, et al., "Alzheimer's Disease, Dental Amalgam, and Mercury," *Journal of the American Dental Association* 130 (1999): 191–99.

178 In 1996 . . . Umeå University in Sweden: A. Berglund and M. Molin, "Mercury Vapor Release from Dental Amalgam in Patients with Symptoms Allegedly Caused by Amalgam Fillings," *European Journal of Oral Science* 104 (1996): 56–63.

178 In 2001 and again in 2008: B. Gottwald, I. Traenckner, J. Kupfer, et al., "'Amalgam Disease'—Poisoning, Allergy, or Psychic Disorder," *International Journal of Hygiene and Environmental Health* 204 (2001): 223–29; J. Bailer, R. Rist, A. Rudolf, et al., "Adverse Health Effects Related to Mercury Exposure from Dental Amalgam Fillings: Toxicological or Psychological Causes?" *Psychological Medicine* 31 (2001): 255–63; D. Melchart, W. Köhler, K. Linde, et al., "Biomonitoring of Mercury in Patients with Complaints Attributed to Dental Amalgam,

Healthy Amalgam Bearers, and Amalgam-Free Subjects: A Diagnostic Study," *Clinical Toxicology* 46 (2008): 133–40.

178 In 1991, researchers from the Medical College of Georgia: J. R. Mackert Jr., M. S. Leffell, D. A. Wagner, et al., "Lymphocyte Levels in Subjects with and Without Amalgam Restorations," *Journal of the American Dental Association* 122 (1991): 49–53.

179 In 2008, American and European researchers: T. A. DeRouen, M. D. Martin, B. G. Leroux, et al., "Neurobehavioral Effects of Dental Amalgam in Children: A Randomized Clinical Trial," *Journal of the American Medical Association* 295 (2006): 1784–92; M. Lauterbach, I. P. Martins, A. Castro-Caldas, et al., "Neurological Outcomes in Children with and Without Amalgam-Related Mercury Exposure: Seven Years of Longitudinal Observations in a Randomized Trial," *Journal of the American Dental Association* 139 (2008): 138–45.

179 In 2008, researchers from Harvard Medical School: D. C. Bellinger, F. Trachtenberg, L. Barregard, et al., "Neuropsychological and Renal Effects of Dental Amalgam in Children: A Randomized Clinical Trial," *Journal of the American Medical Association* 295 (2006): 1775–83; D. C. Bellinger, D. Daniel, F. Trachtenberg, et al., "Dental Amalgam Restorations and Children's Neuropsychological Function: The New England Children's Amalgam Trial," *Environmental Health Perspectives* 115 (2007): 440–46; D. C. Bellinger, F. Trachtenberg, A. Zhang, et al., "Dental Amalgam and Psychosocial Status: The New England Children's Amalgam Trial," *Journal of Dental Research* 87 (2008): 470–74.

179 In 2010, researchers from the University of California: I. Hertz-Picciotto, P. G. Green, L. Delwiche, et al., "Blood Mercury Concentrations in CHARGE Study Children with and Without Autism," *Environmental Health Perspectives* 118 (2010): 161–66.

180 In 2005, researchers from Italy: P. F. Luglie, G. Campus, G. Chessa, et al., "Effect of Amalgam Fillings on the Mercury Concentration in Human Amniotic Fluid," *Archives of Gynecology and Obstetrics* 271 (2005): 138–42.

180 In 2005, researchers from the University of Washington: P. P. Hujoel, M. Lydon-Rochelle, A. M. Bollen, et al., "Mercury Exposure from Dental Filling Placement During Pregnancy and Low Birth Weight Risk," *American Journal of Epidemiology* 161 (2005): 734–40.

180 In 2007, researchers from the University of North Carolina: J. L. Daniels, A. S. Rowland, M. P. Longnecker, et al., "Maternal Dental History, Child's Birth Outcome and Early Cognitive Development," *Paediatric and Perinatal Epidemiology* 21 (2007): 448–57.

180 In 2008, researchers from the University of California: L. Palkovicova, M. Ursinyova, V. Masanova, et al., "Maternal Amalgam Dental Fillings as the Source of Mercury Exposure in Developing Fetus and Newborn," *Journal of Exposure Science and Environmental Epidemiology* 18 (2008): 326–31.

181 "no scientific evidence supports the association": Barrett, "The 'Mercury Toxicity' Scam"; Chey and Buchanan, "Toxins in Everyday Life"; Rathore, Singh, and Pant, "The Dental Amalgam Toxicity Fear."

183 Enter Dr. Oz . . . twenty dollars a bottle: "Toxic Teeth: Are Mercury Fillings Making You Sick?" *The Dr. Oz Show,* https://www.doctoroz.com/ . . . /toxic-teeth-are-mercury-fillings-making-you-sick.

Chapter 17: Vitamin C Doesn't Treat or Prevent Colds

188 Linus Pauling: P. A. Offit, *Pandora's Lab: Seven Stories of Science Gone Wrong* (Washington, DC: National Geographic Press, 2017).

190 "the Linus Pauling Effect": T. Hager, *Force of Nature: The Life of Linus Pauling* (New York: Simon and Schuster, 1995).

190 In 1939, researchers at the University of Minnesota: D. W. Cowan, H. S. Diehl, and A. B. Baker, "Vitamins for the Prevention of Colds," *Journal of the American Medical Association* 120 (1942): 1268–71.

190 Ten years later . . . didn't "have any important effect": D. W. Cowan and H. S. Diehl, "Antihistamic Agents and Ascorbic Acid in the Early Treatment of the Common Cold," *Journal of the American Medical Association* 143 (1950): 421–24.

191 Pauling refused to believe: L. Pauling, "The Significance of the Evidence About Ascorbic Acid and the Common Cold," *Proceedings of the National Academy of Sciences* 68 (1971): 2678–81; T. Goertzel and B. Goertzel, *Linus Pauling: A Life in Science and Politics* (New York: Basic Books, 1995), 202–3.

191 the study to end all studies: A. R. Schwartz, Y. Togo, R. B. Hornick, et al., "Evaluation of the Efficacy of Ascorbic Acid in Prophylaxis of

Induced Rhinovirus 44 Infection in Man," *Journal of Infectious Diseases* 128 (1973): 500–5.

192 Jane Brody . . . non-falsifiable belief: J. E. Brody, "Vitamin C Study Rebuts Pauling," *New York Times*, November 28, 1971.

193 In 1974, researchers at the University of Toronto: T. W. Anderson, D. B. W. Reid, and G. H. Beaton, "Vitamin C and the Common Cold: A Double-Blind Trial," *CMA Journal* 107 (1972): 503–8; T. W. Anderson, G. Suranyi, and G. H. Beaton, "The Effect on Winter Illness of Large Doses of Vitamin C," *CMA Journal* 111 (1974): 31–36.

193 in 1975 . . . researchers reviewed fourteen more studies: M. H. M. Dykes and P. Meier, "Ascorbic Acid and the Common Cold: Evaluation of Its Efficacy and Toxicity," *Journal of the American Medical Association* 231 (1975): 1073–79; T. C. Chalmers, "Effects of Ascorbic Acid on the Common Cold: An Evaluation of Evidence," *American Journal of Medicine* 58 (1975): 532–36.

193 In 1977, researchers in Australia: P. C. Elwood, S. J. Hughes, and A. S. St. Leger, "A Randomized Controlled Trial of the Therapeutic Effect of Vitamin C in the Common Cold," *Practitioner* 218 (1977): 133–47.

193 in 1997, researchers reviewed the results of six more studies: H. Hemilä, "Vitamin C Intake and Susceptibility to the Common Cold," *British Journal of Nutrition* 77 (1997): 59–72.

193 In 2001, researchers at the Australian National University: C. Audera, R. V. Patulny, B. H. Sander, et al., "Mega-Dose Vitamin C in Treatment of the Common Cold: A Randomised Controlled Trial," *Medical Journal of Australia* 175 (2001): 359–62.

194 in 2013, researchers reviewed the results of twenty-nine more studies of vitamin C: H. Hemilä and E. Chalker, "Vitamin C for Preventing and Treating the Common Cold," *Cochrane Systematic Review*, January 31, 2013.

194 Pfizer bought Alacer: S. Daniels, "Pfizer Snaps Up Emergen-C Vitamin Maker," February 29, 2012, nutraingredients-usa.com.

195 researchers at Yale University School of Medicine: E. F. Foxman, J. A. Storer, M. E. Fitzgerald, et al., "Temperature-Dependent Innate Defense Against the Common Cold Virus Limits Viral Replication at Warm Temperature in Mouse Airway Cells," *Proceedings of the National Academy of Sciences* 112 (2015): 827–32.

Chapter 18: Don't Ice Sprains

197 ankle sprains . . . slip on a wet patch: D. Ivins, "Acute Ankle Sprain: An Update," *American Family Physician* 74 (2006): 1714–20.

197 Every day, twenty thousand people: M. P. J. van den Bekerom, P. A. A. Struijs, L. Blankevoort, et al., "What Is the Evidence for Rest, Ice, Compression, and Elevation Therapy in the Treatment of Ankle Sprains in Adults?" *Journal of Athletic Training* 47 (2012): 435–43.

199 In 1989, researchers from New Zealand: E. Laba and M. Roestenburg, "Clinical Evaluation of Ice Therapy for Acute Ankle Sprain," *New Zealand Journal of Physiotherapy* 17 (1989): 7–9.

199 In 2006, researchers from Ireland: C. M. Bleakley, S. M. McDonough, and D. C. MacAuley, "Cryotherapy for Acute Ankle Sprain: A Randomised Controlled Study of Two Different Icing Protocols," *British Journal of Sports Medicine* 40 (2006): 700–5.

199 In 2012, researchers from Amsterdam: van den Bekerom, Struijs, Blankevoort, et al., "What Is the Evidence for Rest, Ice, Compression, and Elevation Therapy in the Treatment of Ankle Sprains in Adults?"

199 In 2013, researchers from Taiwan: C.-Y. Tseng, J.-P. Lee, Y.-S. Tsai, et al., "Topical Cooling (Icing) Delays Recovery from Eccentric Exercise-Induced Muscle Damage," *Journal of Strength and Conditioning Research* 27 (2013): 1354–61.

200 The key to healing is inflammation: J. Robinson, "MOVE on Injury Not RICE," October 18, 2017, https://thischangedmypractice.com/move-on-injury-not-rice.

200 anything that decreases blood flow . . . lengthens the time to healing: J. McCormack, "Mythbuster," *British Medical Journal Talk Medicine*, https://soundcloud.com/bmjpodcasts/mythbuster-on-nsaids-in-sports.

200 "There are no data to show that ice does anything more than block pain": G. Mirkin, "Why Ice Delays Recovery," DrMirkin.com, May 2, 2018, http://www.drmirkin.com/fitness/why-ice-delays-recovery.html; A. Hutchins, "Why Ice Doesn't Help Injury," *Maclean's*, May 20, 2014; A. Harding, "To Ice or Not to Ice a Sprained Ankle? Experts Weigh In," June 8, 2016, https://www.health.com/fitness/ice-sprained-ankle.

201 doctors no longer recommend RICE: S. Byrne, "Why You Should Avoid Ice for a Sprained Ankle," *Consumer Reports*, October 29, 2015; R. Brouhard, "Why You Shouldn't Do RICE for Sprains," *Verywell Health*, September 25, 2018, www.verywellhealth.com/why-you-shouldn't-do-rice-for-sprains-4142771?print.

201 So, what are you supposed to do: Byrne, "Why You Should Avoid Ice for a Sprained Ankle"; Ivins, "Acute Ankle Sprain: An Update."

Chapter 19: Teething Doesn't Cause Fever

202 Hippocrates claimed: L. Guthrie, "Teething," *British Medical Journal* ii (1908): 468.

202 infant deaths in England and Wales: R. S. Illingworth, *The Normal Child: Some Problems of the Early Years and Their Treatment*, 9th ed. (New York: Churchill Livingstone, 1987), 94–101.

203 François Ranchin: A. Tasanen, "General and Local Effects of the Eruption of Deciduous Teeth," *Annales Paediatriae Fenniane* 14 (1968): 1–40.

203 "Teething produces nothing but teeth": R. S. Illingworth, *The Normal Child*, 6th ed. (Edinburgh: Churchill Livingstone, 1975).

203 Teething is a natural process: M. Nemezio, K. M. H. De Oliveira, P. C. Romualdo, et al., "Association Between Fever and Primary Tooth Eruption: A Systematic Review and Meta-Analysis," *International Journal of Clinical Pediatric Dentistry* 10 (2017): 293–98.

203 In 2000, Australian investigators: M. Wake, K. Hesketh, and J. Lucas, "Teething and Tooth Eruption in Infants: A Cohort Study," *Pediatrics* 106 (2000): 1374–79.

204 In 2011, Brazilian investigators: J. Ramos-Jorge, I. A. Pordeus, M. L. Ramos-Jorge, et al., "Prospective Longitudinal Study of Signs and Symptoms Associated with Primary Tooth Eruption," *Pediatrics* 128 (2011): e1–e6.

204 In 2017, Brazilian researchers: Nemezio, De Oliveira, Romualdo, et al., "Association Between Fever and Primary Tooth Eruption."

Epilogue: Choosing Wisely

205 In 2017, *The Atlantic* published an article: D. Epstein, "When Evidence Says No, But Doctors Say Yes," *The Atlantic*, February 22, 2017.

205 op-ed in the *New York Times*: A. E. Carroll, "It's Hard for Doctors to Unlearn Things. That's Costly for All of Us," *New York Times*, September 10, 2018.

206 "It's hard to get people in any profession to do the right thing": Epstein, "When Evidence Says No."

206 "Things have gotten better": Quoted in Epstein, "When Evidence Says No."

206 Regarding inappropriate testing: S. Brownlee, "Getting Physicians to Stop Delivering Low-Value Services," *Health Affairs Blog*, October 11, 2018, https://www.healthaffairs.org/do/10.1377/hblog20181003.943891/full.

206 After evaluating a woman who wanted an operation: Quoted in Epstein, "When Evidence Says No."

207 relationship between patient ratings and receipt of an antibiotic prescription: K. A. Martinez, M. Rood, N. Jhangiani, et al., "Association Between Antibiotic Prescribing for Respiratory Tract Infections and Patient Satisfaction in Direct-to-Consumer Telemedicine," *Journal of the American Medical Association Internal Medicine* 178 (2018): 1558–60.

207 "Calling patients consumers": Brownlee, "Getting Physicians to Stop Delivering Low-Value Services."

208 "Sometimes investigator beliefs in scientific circles": A. Tatsioni, N. G. Bonitsis, and J. P. A. Ioannidis, "Persistence of Contradicted Claims in the Medical Literature," *Journal of the American Medical Association* 298 (2007): 2517–26.

208 ORBITA: R. Al-Lamee, D. Thompson, H.-M. Dehbi, et al., "Percutaneous Coronary Intervention in Stable Angina (ORBITA): A Double-Blind, Randomised Controlled Trial," *Lancet* 391 (2018): 31–40.

209 "I would like to thank the patients participating in #ORBITA": Epstein, "When Evidence Says No"; Brownlee, "Getting Physicians to Stop Delivering Low-Value Services."

209 "2.4 million unnecessary operations": P. F. Stahel, T. F. Vander Heiden, and F. J. Kim, "Why Do Surgeons Continue to Perform Unnecessary Surgery?" *Patient Safety in Surgery* 11 (2017): 1.

210 spending on the marketing of drugs: L. M. Schwartz and S. Woloshin, "Medical Marketing in the United States: 1997–2016," *Journal of the American Medical Association* 321 (2019): 80–96.

210 "As a pharmacy student . . . She wanted those medicines": H. Bodenstab, personal communication, May 2019.

211 Judith Haimes, a self-proclaimed psychic: M. Digirolamo, "Judge Throws Out $986,000 Jury Award to Psychic," UPI, August 9, 1986.

Index

About the Author

PAUL A. OFFIT, MD, is the director of the Vaccine Education Center and attending physician in the Division of Infectious Diseases at the Children's Hospital of Philadelphia, as well as the Maurice R. Hilleman Professor of Vaccinology and a professor of pediatrics at the Perelman School of Medicine at the University of Pennsylvania. He is also the co-inventor of the rotavirus vaccine, RotaTeq, recommended for universal use in infants by the CDC in 2006 and by the WHO in 2013; for this achievement, Dr. Offit was honored by Bill and Melinda Gates during the launch of their foundation's Living Proof Project for global health. He is a recipient of many awards, including the Young Investigator Award in Vaccine Development from the Infectious Disease Society of America, a Research Career Development Award from the National Institutes of Health, the President's Certificate for Outstanding Service from the American Academy of Pediatrics, the David E. Rogers Award from the American Association of Medical Colleges, the Odyssey Award from the Center for Medicine in the Public Interest, the Maxwell Finland Award for Outstanding Scientific Achievement from the National Foundation for Infectious Diseases, the Distinguished Alumnus Award from the University of Maryland School of Medicine, the Porter Prize from the University of Pittsburgh School of Public Health, the Jonathan E. Rhoads Medal for Distinguished Service to Medicine from the American Philosophical Society, the Gold Medal from the Sabin Vaccine Institute, the John P. McGovern Award from the American Medical Writers Association, and the Geoffrey Beene Foundation Builders of Science Award from Research/America. In 2011 Dr. Offit was elected to the National Academy of Medicine and in 2015 to the American Academy of Arts and Sciences.